SEARCHING FOR THE
SOROR MYSTICA

The Lives and Science of Women Alchemists

Robin L. Gordon

University Press of America,® Inc.
Lanham · Boulder · New York · Toronto · Plymouth, UK

Copyright © 2013 by
University Press of America,® Inc.
4501 Forbes Boulevard
Suite 200
Lanham, Maryland 20706
UPA Acquisitions Department (301) 459-3366

10 Thornbury Road
Plymouth PL6 7PP
United Kingdom

British Library Cataloging in Publication Information Available

Library of Congress Control Number: 2012952642
ISBN: 978-0-7618-6055-6 (paperback : alk. paper)

Carrington, Leonora (1917-2011) © ARS, NY. The Chrysopeia of Mary the Jewess, 1964. Oil on canvas. Canvas: 59 x 35 7/16 in. (149.86 x 89.9922 cm). © Leonora Carrington/Artists Rights Society (ARS), New York. Cover photo courtesy of a private collection.

Dedication

To the many women throughout time who have sought to understand the universe.

Contents

List of Tables and Figures

Foreword

I have been waiting for this book on the history of female Alchemists to be written for at least forty years. How I wish it had existed when I did my own research back in the seventies on the Women of Surrealism, on artists such as Leonora Carrington, whose painting THE CHRYSOPEIA OF MARY THE JEWESS, one of the early female Alchemists, found its way, magically, to the cover of this book. Robin Gordon's work has its own unique alchemical formula, its own "Chrysopeia."

Her "work" (in the alchemical sense of the term) restores to history the lineage of women alchemists that had been excluded from mainstream accounts of science until now. Gordon accomplishes this by combining the essential elements of a contemporary feminist analysis with an understanding of how the interconnectedness of various fields such as Jewish mysticism, Christianity, Astrology, Jungian Psychology, Magic, and the Healing arts is absolutely essential to grasping the full range of disciplines in which these female Alchemists had expertise. They transformed the patriarchal paradigm based on dualism and separation into a feminist paradigm stressing the holistic interrelationships of all aspects of spirit and nature. Both Robin and I were stunned to find that the process of doing research on alchemy actually set in motion a chain reaction akin to what was meant by transformation in our own lives. Unexpected encounters resulted that illuminated our individual quests. These women who explored the alchemical process in the past were the "wise women" of history, not witches, but Alchemists and Scientists. To read Robin Gordon's account of their lives and of her own process enables us to make this important encounter with the visionary female wisdom figures she refers to as the *Soror Mystica*.

Gloria Feman Orenstein, Ph.D.
Professor Comparative Literature and Gender Studies
University of Southern California, Los Angeles
July 2012

Preface

Typically, I do not like to read a preface. I want to get into the book right away, skipping all the things the author thinks she or he should tell me before I begin reading *the good stuff*. However, I would like to tell the reader the story of this book. It has been eight years in the researching, collecting, and writing. The book was nearly complete two years ago when I lost my bearings in the writing. The reaction I was getting from potential publishers caused me to question if my work was too historical or too Jungian. A prominent Jungian writer wrote that it was not Jungian enough. Some historians of science told me that Jung didn't have much to offer about the study of alchemy. I became convinced that in order to publish the stories of these women alchemists, I would need to revise. That simple statement began a two year inner and outer journey that delayed progress but was very much needed. Jungians like to speak of allowing ideas to cook and my cooking time was akin to a slow roast.

Sir Ken Robinson quoted Abraham Lincoln in the technology talk he gave at the February 2010 TED Conference. He stated:

> I came across a great quote recently from Abraham Lincoln . . . . He said this in December 1862 to the second annual meeting of Congress. . . . 'The dogmas of the quiet past are inadequate to the stormy present. The occasion is piled high with difficulty, and we must rise with the occasion.' [Robinson adds] I love that. Not rise to it, rise with it. [Lincoln quote continues] 'As our case is new, so we must think anew and act anew. We must disenthrall ourselves, and then we shall save our country.'
>
> [Robinson] I love that word, 'disenthrall.' You know what it means? That there are ideas that all of us are enthralled to, which we simply take for granted as the natural order of things, the way things are. And many of our ideas have been formed, not to meet the circumstances of this century, but to cope with the circumstances of previous centuries. But our minds are still hypnotized by them, and we have to disenthrall ourselves of some of them. Now, doing this is easier said than done. It's very hard to know, by the way, what it is you take for granted. And the reason is that you take it for granted.

Writing a book that crosses disciplines is not so difficult for the author but can present quite a challenge for the reader. When we choose to spend time with an author, we usually begin with the genre that interests us most. In the case of non-fiction, we focus on a subject such as history, psychology, or biography. The choices are vast. What then to do with a book that includes elements of many subjects?

Upon listening to Robinson's talk, it dawned on me that I needed to *disenthrall* myself from thinking that the women alchemists' book had to be either this or that. It was the artificial dichotomy I had created that needed consciousness and revisioning. The American Heritage Dictionary (2000) defines disenthrall as "to free from a controlling force or influence" (p. 517). The word traces to Middle English when enthrall meant "to enslave or to put into bondage" (p. 596). I had to *disenthrall* my thinking that the book needed to be a historical analysis or in a different iteration, a focused plunge into the complexities of Depth Psychology. Why could it not be both and more besides? The stories of the women alchemists are grounded in history; however, they are incomplete without mentioning magic, occult beliefs, Jewish mysticism, Christian apocalyptic study and what these ideas meant to the people involved. Recounting a historical timeline of events grounds the story in context. Examining how the participants thought about their world that was not the linear place we think exists in modernity, is critical as well.

My conclusion is to inform the readers about what to expect and then give them the space to make sense of the story. That is what you will find in this book. There is history of science, biography, classical Jungian psychology, women's studies, theology and a dash of the occult sciences. I hope you will see that the early scientists or natural philosophers did not separate subjects from each other the way modern academics tend to do. They were interested in how the universe worked and that meant studying everything, from astrology and physics to Jewish mysticism and the Christian Bible. They constructed connections that the modern thinker might overlook or more deliberately, dismiss as preposterous. In the interest of avoiding the writing of an encyclopedia, I have chosen to examine a few elements from these various disciplines but the reader is encouraged to do further exploration on topics that are particularly resonant with your interests. Thus, the reader will find many types of stories in this account of women practicing alchemy. Diverse subjects will be included that range from politics, religion, scientific inquiry, and even the way love can result in some misguided choices.

Finally, I am a teacher and thus, am prone to pose questions without providing the reader with the so-called *right answer*. Some authors feel obligated to give the reader the answers to all the questions that emerge from their work. However, my hope is that I have left space for readers to have questions emerge that compel one to sit in a comfy chair with a cup of tea or coffee, staring out a window, and to just—think.

Robin L. Gordon, Ph.D.
Arcadia, California
February 3, 2010/April 12, 2012

Acknowledgments

I would like to acknowledge the following libraries and more specifically, the many librarians who provided resources, writing spaces, and guidance when needed. I think there is a special place in the universe for librarians who guard the collective knowledge of our planet, most often without recognition. They have often been the brave guardians of the written word when confronted by tyranny.

Huntington Library, San Marino, California

Bibliotheca Philosophica Hermetica—J.R. Ritman Library, Amsterdam, Netherlands

British Library, London

Wellcome Library, London

Westminster Abbey Library, London

Additionally, I would like to thank Samantha Kirk from University Press who originally believed in this book, Laura Espinoza, current Editor at University Press of America for her guidance and wonderful feedback, and Megan Barnett who has the editing eyes of a hawk! The index was generated via texyz.com.

Other contributors to this book include: Christy Ann Roach (translator for Marie Meurdrac's *La Chymie*), Daniel Roach (further translation work), Gloria Avrech and Gloria Orenstein—alpha and omega—*maggids* both, Ilene Susan Fort (LACMA), Chelsea Rhadigan (Artists Rights Society), Diana Reeve (Art Resource Inc.) and the very generous owner of *The Chrysopeia of Mary the Jewess* who made the book cover possible.

And lastly, thanks must be given to my family members and friends who for eight years have had the patience to listen to me talk about these women and their work.

Chapter 1

Introduction: Who Are the Women Alchemists?

When I began this small treatise, it was for my sole satisfaction, so as not to lose memory of the knowledge that I had acquired by means of long toil, and by divers experiments repeated several times. I cannot conceal that seeing it achieved beyond what I had dared to expect, I was tempted to publish it; but if I had reason to bring it to light, I had even more reason to keep it hidden and not to expose it to general censure. . . . from the foreword of *La Chymie charitable et facile, en faveur des Dames Charitable* [*Easy Chemistry for Women*], 1656 by Marie Meurdrac.[1]

Picture, if you will, the workspace of an alchemist from the seventeenth century. There are oddly shaped jars scattered about the workroom that contain samples of minerals and herbs. A fire burns gently under a brick oven, maintaining a constant temperature. The alchemist is stirring a mixture of ingredients that has been soaking in human urine for many days and is now ready to be distilled. There are a few precious books stacked in the corner of the room that describe hundreds of sets of directions on how to do the work. It is quiet in the workroom today and the alchemist, noticing that the mixture is nearly ready, brushes the hair out of her eyes and peers at the fire to be sure it is burning at the right heat for the next step in the work.

Who is the *soror mystica*, the mystic sister, the female practitioner of the ancient art and science of alchemy? Furthermore, why would anyone in the 21st century care to understand such an ancient science? What is there in the alchemical tradition that could possibly speak to our modern era when even pre-kindergarten children are adept with technology? In this book, I will focus on women whose stories in alchemy have been neglected until recently, but are emerging in contemporary research on women's early work in science. I have reorganized the picture of their work to show why I believe they can be called alchemists, despite resistance from some historians. I will also show the reader how the philosophical foundations of alchemy are still quite relevant and active in modern society. The history of intertwining alchemical beliefs and religion, as well as the effects on politics throughout Western history will resonate with the contemporary reader. I anticipate that the reader will also recognize similarities between the theo-politics that resulted in the execution of Charles I in 1649 and the current issues of separation of church and state that challenge the modern United States. Finally, I will show how the psychological aspects of alchemy are still in place today, despite its ancient and seemingly outmoded tradition.

The art of alchemy was referred to by some alchemists as the *Magnum Opus* or *Great Work*. Aristotle had taught that matter was created from the four elements: earth, air, fire, and water. Earth was described by the *dominant and subordinate* characteristics of Dry and Cold. Fire was Hot and Dry; Water was Cold and Moist; Air was Moist and Hot. Changing the second or subordinate characteristic allowed the elements to be converted into each other. Earth that is dry and cold is changed to fire by converting the cold to heat and so forth.[2]

It was believed that various forms of matter contained different proportions of the elements and could be transmuted from one to another by changing proportions.[3] Gold was considered a perfect form of matter with silver a close second. The elevation of gold and silver to the highest forms of matter was linked to mankind's reverence for gold, probably due to its association to the sun, giver of life.[4] People depend upon the sun as the source of food, warmth, and growing things. The sun is a holy entity as well as the moon, the keeper of monthly cycles. Thus, of course, gold and silver would become revered as symbols of these two most comforting and life sustaining presences. Alchemical work was all about re-creating a natural process of generating precious metals such as gold, by disassembling imperfect matter (lead or some other metal considered less perfect), rearranging the proportions of the elements, and reassembling the substance to produce gold or silver. The process for creating healing elixirs was similar and both types of work will be discussed in this book.

The tradition of women practicing alchemy, as well as numerous other scientific endeavors, can be found occurring from before recorded history through early antiquity. Keep in mind that the scientific disciplines were not separated as they are today. One could study astronomy as well as physics and biology. The world was seen as a reflection of the Creator and to understand the natural world was to know God. It becomes clear that in order to examine the history of alchemy, we need to look further than the traditional histories of science and chemistry where authors discuss early chemists and how they moved the field away from its spiritual-alchemical roots, to one that focused on the rational Scientific Method that emerged in the seventeenth century. When one reads early writings in areas such as agriculture, medicine, and healing, one hears the stories of less well-known female practitioners who were carrying out the same fledgling chemistry as their more famous brethren.

Margaret Alic argues that many of the goddesses who were associated with healing and agriculture such as Isis, Athena, Demeter, Hygeia, and Ishtar were actually living women who had gained a reputation for their work.[5] Her point is that women have been studying science for a very long time. Some of the earliest written records of scientific work can be traced to circa 2278-2263 BCE, at the time the pyramids were built in Egypt. There is substantiation of Mesopotamian women distilling perfume in 1200 BCE.[6] There is sufficient evidence of women practicing alchemy in China to warrant further research beyond the scope of this book but it goes back as far as the first century BCE.

Even in antiquity, numerous women were educated in medicine and attended schools in Sais and Heliopolis, ancient Egyptian cites. An inscription from the Temple of Sais reads, "'I have come from the school of medicine at Heliopolis, and have studied at the woman's school at Sais where the divine mothers have taught me how to cure disease.'"[7] It is also believed that Moses studied medicine at Heliopolis along with his wife, Zipporah, circa 1500 BCE.

The famous mathematician, Pythagoras of Samos c. 582-500 BCE, is said to have developed his "moral doctrines"[8] under the tutelage of a priestess in Delphi, Themistoclea. He founded the Pythagorean Community also known as the "Brotherhood"[9] circa 540-520 BCE. The community consisted of female as well as male teachers. The Pythagorean theory of the universe consisted of a universe that was ordered into spheres, the most perfect mathematical shape. The planets, moons and sun orbited the earth following a spherical path. The orbits of the planets or spheres were thought to be spaced in precise mathematical intervals, just like those found in the musical scale. The Brotherhood believed that mathematics, and concomitantly music, was a pathway to understanding the Divine. Pythagoras was believed to have married a young female student, Theano, in his old age. Theano was a scholar, wrote on many subjects such as the Pythagorean Golden Mean, and carried on his work after Pythagoras was murdered and the school was dispersed.

The *Hetairai* are another example of the less-than-traditional education achieved by some women in antiquity. Prostitutes or courtesans of Greek men were not bound by the same restrictions as their wives and were often educated. They were called *hetairai*, which translates as a companion to men.[10] Algaonice from Thessaly (5th century BCE) is an example of an educated woman from Roman antiquity who could predict solar and lunar eclipses. However, on the other hand, we have Aristotle claiming that women were deformed men and that the soul was located in male semen. It is no wonder that misogynist beliefs prevailed in light of the great influence Aristotle had on scientific thinking for generations.

A number of women are identified as practicing medicine in Medieval England.[11] Matilda from Wallingford, Berkshire is thought to be one of the earliest recorded female physicians. She practiced circa 1232 CE and was known as "Matilda la Leche"[12] as well as "sage femme." [13] Agnes 'medica' from Huntingdonshire, circa 1270 CE, was given that nickname by her neighbors as a compliment to her medical abilities.[14] A female surgeon (c. 1286), from London was said to have learned the craft from her father and brother. All that is known of her identity is that she was named Katherine but was called, "'la surgiene.'"[15] What is clear is that despite the patriarchal oppression of women during early history, some women still found a way to become educated and practice their work.

How the Search for the Soror Mystica Began

Initially, my work began with previous research I had completed in which I examined the psycho-historical significance of the philosopher, Baruch Spinoza, and his relationship with alchemy.[16] I argued that the study of alchemy played a considerable role in developing major scientific theories, a notion supported by research on the history of science. Alchemical study consists of an intertwined relationship between philosophy, theology, matter, spirit, and soul. There is evidence that the emergence of a *separatio*,[17] a divergence in people's understanding of the relationship between the ideas of body, soul, and spirit, occurred around the seventeenth century, at the time of the Scientific Revolution; yet, I could not reconcile how that split could take place within the context of an alchemical paradigm that was held by most philosophers of the time such as Boyle, Newton, and numerous other scientists who studied alchemy. Alchemy stated that body, soul, and spirit were bound together and separating them from each other, a step needed in order to finally create the Philosopher's Stone, was a difficult task fraught with error. In the midst of researching these well-known scientists, I also began noticing tantalizing morsels of information concerning what I came to think of as *the ladies of alchemy and science*. It seemed that there were many women who studied alchemy and were sometimes referred to as a *soror mystica* (mystic sister).

Investigating the story of the *soror mystica* leads the researcher down disparate paths. The term *soror mystica* usually refers to the female helper of the alchemist. For example, in *Psychology and Alchemy*, the noted depth psychologist Carl Jung, identified a young woman named Theosebeia as a *soror mystica* and the helpmate of Zosimos of Panopolis.[18] Possibly the actual sister of Zosimos as well, Theosebeia assisted him in writing one of the first alchemical encyclopedias, *Cheirokmeta* c. 300 CE.[19] The encyclopedia consisted of 28 books and included references to the work of both Maria Prophitissa and Cleopatra (another alchemical investigator, not necessarily the well-known queen).

Another source of information on the *soror mystica* is the *Mutus Liber* (1677). This work consists of a series of 15 engravings that illustrate the steps in accomplishing alchemical work. The author is unknown except by the name, "'Altus—the high, deep, or profound one."[20] The *Mutus Liber* is unusual in that it depicts the alchemist working alongside a woman, possibly his wife, although the term, *soror mystica* is not used in the treatise. However, besides the women pictured in assorted woodcuts and engravings in the alchemical literature such as in the *Mutus Liber*, as well as references to the work of Maria Prophitissa, I initially found very little evidence of female alchemists.[21]

A few unfamiliar names, however, did emerge in my research on the aforementioned men. For example, Lady Katherine Ranelagh (1614-1691), sister of the famous scientist, Robert Boyle, opened her home to her brother and his alchemical colleagues. Perhaps I projected my own scientific curiosity onto her, but I could not help but think that someone who was associated with the stories of their research in natural philosophy would surely be involved with the work itself. There is evidence of her work in what some historians have described as medical chemistry or iatrochemistry. Katherine also studied the *Kabbalah*, the body of Jewish mysticism, and I wondered if she linked this religious learning to understanding the nature of the Philosopher's Stone. Does the dearth of written evidence that she worked as an alchemist herself mean that she acted only as an assistant for her notable brother?

Months of combing through the collection at the Huntington Library in San Marino, California, proved fruitful in my search for these women practitioners. For example, Marie Meurdrac was described by historian, Lucia Tosi as the first woman to publish a book on alchemy or early chemistry in *La Chymie charitable et facile, en faveur des Dames Charitable* [*Easy Chemistry for Women*].[22] In addition to providing detailed instructions for the creation of medicines and cosmetic ointments, Marie exhorted her readers to follow her example and distribute these remedies free of charge to the poor, a practice that I eventually learned was common for many of the women alchemists. Marie also offered to teach women in her own laboratory if they felt unsure about attempting the alchemical work on their own.

The stories of women who studied the natural world have been given greater examination as described elegantly by writers such as Margaret Alic, Lynette Hunter, Sarah Hutton, Merry Weisner and Tara Nummedal.[23] Certainly, there were fewer educated women than men prior to the twentieth century, but many women, nonetheless, found a way to challenge their minds and immerse themselves in a sincere study of the natural world. Thus, intrigued and invigorated, I continued to look for these elusive sisters in science. I have since compiled the names and stories of many women who were both skillful alchemists as well as researchers in several fields of science. I will describe the numerous ways they manifested their practice that was not always obvious and has been largely disregarded in traditional, alchemical literature.

Regarding research method, I employ the hermeneutic and heuristic methods, ones that are often employed in the field of social science and as well, depth psychology. Hermeneutics provides a framework for examining a question from several perspectives.[24] Unlike a straightforward answer to one question, hermeneutics provides a space for exploring the numerous questions that continue to emerge in the course of research. One question leads to another and the resulting work may be quite different than what the researcher first planned. For example, I had not anticipated the significant role played by the study of the Christian Apocalypse that I found in the stories of many alchemists. The reader will find in chapter seven that one cannot disregard that area of study as peripheral. For some of the women in this book, their theology was so connected to their work that the question emerged and had to be addressed.

Another significant element of hermeneutic science will be seen throughout this book. Clark Moustakas describes this type of analysis; "hermeneutic science involves the art of reading a text so that the intention and meaning behind appearances are fully understood."[25]

Heuristics, as developed by Clark Moustakas, recognizes that the researcher is intimately involved with the subject of the research. There is some disagreement in the research community on how to define heuristics but the simplest way to explain the way I have used it in this book is to state that relating the stories of the women alchemists is one goal but I will also discuss what their stories mean to me. Moustakas explains:

> It refers to a process of internal search through which one discovers the nature and meaning of experience and develops methods and procedures for further investigation and analysis. The self of the researcher is present throughout the process and, while understanding the phenomenon with increasing depth, the researcher also experiences growing self-awareness and self-knowledge. Heuristic processes incorporate creative self-processes and self-discoveries.[26]

My work is subjective but I have tried to present alternative views to provide balance; the readers are invited to come to their own conclusions.

Finally, I have similarly approached my research from what is called a psycho historical perspective. Edward Edinger writes, "Everything that happens in the psyche happens for an adequate reason."[27] The field of depth psychology, which includes the study of the role of the unconscious in our psychic development, is the framework I will use to discuss my interpretation of what I believe occurred in the lives of the women alchemists. It will help the reader to understand that depth psychologists study the whole psyche, the unconscious as well as the conscious ego. Jung was followed by brilliant thinkers such as Marie Louise von Franz and Edward Edinger, who prodded the field further into explorations of how the unconscious is present and active in our daily lives. James Hillman added an archetypal element to depth psychology, characterizing the universe as alive and interconnected.[28]

Depth psychology has a vested interest in traveling back in time to examine historical events through its particular lens. Our culture's history is a story of our holistic development, the inner is reflected in the outer, or as the old alchemists often quoted, "as above, so below." History does not unfold disconnected from psyche; it is a reflection, and often an unconscious one, of psyche's development. Alchemical practices did not develop independently of the culture of the practitioners. It was spoken of in the terms of each generation, often employing its unique religious symbolism be it pagan, Gnostic, Christian, or of the Kabbalah. Carl G. Jung illuminated the analogy of alchemy and psychological development. It is within that context that I plan to reexamine events that are elements lacking in the story of Western evolution and the Scientific Revolution.

Another important aspect of understanding the *idea* of the psyche is that it has both a feminine and masculine nature, regardless of one's sex. Balancing the feminine and masculine principles is a goal of individuation or psychic development. Eastern philosophy speaks of the feminine *yin* and the masculine *yang* as being active principles in our psyches. It can be argued that macro-entities such as culture and world consciousness contain these opposing aspects as well. The alchemical term *coniunctio*[29] describes the joining of the two principles, resulting in something that is greater than the sum of the parts. This challenge of achieving harmony with the delicate balance of differing psychic energies will be illuminated in the course of telling the women alchemists' stories.

Keeping in mind the existence of the masculine and feminine principles of psyche, a curious dichotomy appeared when I began my investigation searching for women alchemists. I sensed my inner psychic feminine principle pointing me strongly in the direction I should follow, via the tantalizing emergence of numerous names of women who appeared to be associated with alchemical work. The outer masculine; however, literally discounted these women. In one instance, I was informed via email by a prodigious author of traditional alchemical literature (physical alchemy) that in each case I cited in my email to him, the woman was fictional, inadequately documented, associated with herbal remedies which he stated was not "true alchemy," or her alchemical status was the product of wishful thinking on the part of Jungians who wish to see *coniunctio* everywhere whether it exists or not. As I strive to be both a careful scholar and imaginative thinker, I felt that his points needed to be researched further. Yet, I also believed that these women, at a psychic level, were admonishing me that they were not phantasmagoria and furthermore, they cared little whether or not their work had been well documented by academics.

The deeper connection between the masculine and feminine principles of psyche is at the heart of my work. Historical accuracy regarding women's practice of alchemy and what it looked like is important to me, not for the academic, scholarly, ego-oriented stamp of approval, but for the very fact that there is even a question of whether or not they studied the science. I question why the field of depth psychology, which values the feminine psyche, offers so little information regarding these women; yet, recounts so many stories of their male counterparts? Reading numerous accounts of alchemists results in a long list of male practitioners but only a handful of women. Why would women *not* have practiced this early chemistry? Why would certain scholars be so sure they did not? Why does one have to *prove* women studied alchemy as opposed to simply accepting the logic that they must have, considering their other well-known pursuits in astronomy, natural philosophy, and mathematics?

Therefore, I will also examine the nature of what some writers disparagingly label *women's alchemy*, as if it were of lesser value, less meaningful, not sufficient. We do have examples of women's early medicinal work in the rare, surviving copies of their recipe books or "receipts." These women boiled herbs, made poultices, and processed curative food using the identical operations and

implements employed by the alchemists.[30] Alchemical practices such as distillation were, in fact, commonly used by non-alchemists as well as the traditional
alchemist that comes to mind when we imagine some fellow working in his laboratory.

The notion that alchemy could manifest in different ways seems clear to me
and is a central argument in my work. Robert Multhauf argues that telling the
story of chemistry, the *child* of alchemy, necessitates examining how medicine
and chemistry are completed by each other.[31] This entails embracing alchemy as
a legitimate science, rather than pretending that the work was an uncommon,
occult practice. Margaret Alic discusses the role of alchemy in the manufacture
of perfumes and cosmetics, and acknowledges that, "the work of the early alchemists was sometimes called *opus mulierum*—'women's work'"[32] thus dimin-
ishing alchemy's importance. Lynn Thorndike, however, quotes Libavius, a
German alchemist who wrote *Neo-Paracelsica* (1594) and *Alchymia* (1597). He
defined alchemy as "the art of accomplishing masteries and extracting pure es-
sences from compounds by separating the body, while Chymia or chemistry was
the second part of Alchymia and concerned with making chemical species."[33]
Thus, alchemy subsumes chemistry rather than vice versa.

W.S.C. Copeman examines the connection between alchemy and the development of the field of medicine. He states, "No learned physician could afford
to be without a working knowledge both of alchemy and astrology."[34] Alchemy
and astrology were the foundations for studying the nature of matter and thus,
contributed to the development of medical practice. Copeman describes how
Queen Elizabeth learned chemistry from her personal astrologer and alchemist,
Dr. John Dee.[35] Even Pope John XXI is described as having been a physician
and utilizing a laboratory at his Palace in Avignon where he also experimented
with alchemy. Copeman points out that subsequent to the Pope's failure at
achieving transmutation, he issued his Papal Bull declaring that alchemy was not
acceptable to God.[36] Consequently, these records of men practicing alchemy in
conjunction with healing make the use of the term, *women's alchemy*, puzzling.

Organization of Searching For the Soror Mystica by Creating A Way to Look At Alchemy

In order to develop a framework for looking at the work of women alchemists, I
used a method that is common in qualitative research in which patterns in observed events or of participants' responses are discerned by the researcher. I
began by asking the question, why did alchemists, both men and women, study
alchemy? I distilled my findings into categories. (See Table 1.1)

I concluded that there were four major categories of alchemists. There were
the Natural Philosophers who would go on to become modern scientists. They
sought to understand the workings of nature and thus, God. Creating the Philosopher's Stone was in service to that understanding. The Spiritualists' goal was

inner transformation. They believed that their alchemical work resulted in both physical transmutation as well as, and more importantly, inner transformation. The Profiteers were usually blatant charlatans. Their goal was profit and they had numerous tricks for appearing to create gold when in fact, they did nothing of the sort. However, some in this category truly believed that transmutation was possible and that vast riches awaited the successful alchemist and/or his sponsor. Finally, I have titled the category that was focused on healing as the Paracelsians. They, like Paracelsus, believed in creation of the Philosopher's Stone which could be used to create miraculous healing medicines.

Table 1.1 Analysis of the Various Goals of the Alchemists

Patterns that indicate how alchemy affected the alchemists' lives. This is important in understanding the development of the Western psyche and the effects on worldview/ cosmology / science.

Spiritualists	**Natural Philoso-phers (Scientists)**	**Profiteers**	**Paracelsians**
Goal: Creating the Philosopher's Stone or Elixir of Life but with the goal of inner trans-formation. ∞ **Elaboration:** French writes that, "Alchemy was by no means limited to the art of trying to transmute base metal into gold; in fact, its essential function was to transmute the hu-man spirit through gnosis."[37]	**Goal:** Obtaining knowledge in all areas of natural philosophy and the creation of the Philosopher's Stone is attempted in service of learn-ing how the world works. ∞ **Elaboration:** The underlying goal is to understand the world and thus, God. How they did this depended upon their framework of thinking about alchemy.	**Goal:** Ultimately to make gold for their own profit and/or for that of someone in power. ∞ **Elaboration:** We acknowledge that there were many so-called alche-mists who sought gold and material wealth, but the *true alchemists* charac-terized those indi-viduals as charla-tans.	**Goal:** Could be the creation of the Philosopher's Stone or Elixir of Life but the focus is on healing. ∞ **Elaboration:** Para-celsus is the most notable example.

It became clear to me while reading sixteenth and seventeenth century reci-pe books that the directions for creating medicines were identical with tradition-al alchemical instructions for making gold that have been recorded by known alchemists. Women were, of course, practicing alchemy. Tara Nummedal asserts that the lack of traditional evidence—that is, the dearth of writing by women alchemists—has misled historians to believe that few, if any, women practiced alchemy. She argues:

> If, however, we shift our focus beyond printed alchemical texts and toward the
> archival sources characteristic of the new social history, suddenly we do find
> women alchemists. Letters, contracts and criminal trial dossiers reveal that both
> noble and common women engaged in the patronage, theory and practice of al-
> chemy.[38]

I have indeed found that an expansion of focus to include archival research has
been extremely fruitful and illuminating. For example, reading the letters exchanged between Katherine Boyle and her brother, Robert, allows their personal
voices to be heard and gives deeper insight into their work. One can see clearly
that Katherine understood what was needed in a laboratory and the processes of
the way alchemical operations worked.

Another example of the benefits of archival research is illustrated by reading women's diaries.[39] Women's diaries and memoirs began to be published
with much greater frequency in the 1600s.[40] Granted, most of the writing origi-
nated with upper and middle class women who could both read and write, although their work was not always met with approval. For example, the great
diarist, Samuel Pepys, upon hearing his wife, Elizabeth's, written description of
her life with him, furiously ordered her to tear up the insulting piece.[41] Women's
published autobiographies were meant to be read by the public and were edited
as necessary, thus, often lack details that might have caused embarrassment for
the family or might suggest that the writer held heretical beliefs. Many women
went to great pains to keep their diaries private. Some used indecipherable
shorthand so that after their death, the contents of their writing would remain
vague. Several families destroyed journals that they deemed to be unfair to
them.

A number of women just wanted their writing to be secret until their death
but if they were really worried, destroyed their work when they felt death might
be imminent. However, personal memoirs, commentaries on women's personal
reading tastes, and inclusion of household notes in their journals describe the
sometimes boring but always enlightening goings-on in the household. Many
upper class women were intimately involved in the oversight of their homes and
often participated in actual chores being accomplished on their estates.
Women's reflections on religious themes, recording fears and joys surrounding
marriage, childbirth, and child rearing give an intimate view of their lives. In
reading these accounts, we do find evidence of alchemical knowledge and practice.

Women also wrote more formally during the years 1600–1700 CE although
with less frequency than men.[42] Keeping in mind that English printed books
date from the 1400s, and that women did not write anywhere on the scale of
men, it is not surprising that there were very few published works during the
time covered by my research. Furthermore, Patricia Crawford notes that some
publications were first credited to men. For example, William Leibniz's 1690
work, *Opuscula Philosophica* (1690) is more accurately based on Anne Con-

way's work. Franciscus Mercurius Van Helmont (son of John Baptiste, another famous alchemist) was given credit for Conway's 1692 manuscript, *The Principles of the Most Ancient and Modern Philosophy,*[43] but what he really did was translate her work into Latin.[44] In the seventeenth century, only 231 women were responsible for the 653 books published in England. Their commitment to writing is laudable considering that by doing so, they stepped outside their accepted feminine role, despite feeling intimidated by men's often superior education and being subject to criticism from those in society who disapproved of their work.

Lynette Hunter and Sara Hutton also examine why it has been so difficult to document women's contributions to alchemy, as well as to science in general. They remind us that in the seventeenth century, when the scientific method was in its infancy, much of what women were doing in the context of early chemistry, which I am arguing should really be labeled alchemy, was accomplished in their kitchens. If the patriarchy, represented by the scientific community was to validate that work, the alchemists had to move from the kitchen to a laboratory.[45] Certainly, gentlemen scientists were not going to create something as important as the Philosopher's Stone in the kitchen! Furthermore, by elevating the role of the gentleman scientist, many women came to be viewed as nothing more than helpmates. This was not necessarily done mean spiritedly but certainly reads condescendingly. They might even be applauded for their support such as Katherine Boyle Jones, Lady Ranelagh but recognition of their work in the sciences is overlooked since it was not published as such. Sir Alan Cook writes as late as 1997 in the *Notes and Records of the Royal Society*, "Catherine [Barton, Isaac Newton's niece], like Lady Ranelagh and Lady Masham [daughter of Ralph Cudworth, philosopher], advanced the scientific revolution by her support of a great man, the principal revolutionary in fact."[46]

The Women's Voices

Carol Gilligan's seminal work, *In a Different Voice* addresses Western society's propensity to define cultural mores using a masculine model of behavior as the norm.[47] She explains how behavior outside the norm gets labeled as deviant or abnormal. Thus, if women accomplished goals in a different way than did men, rather than viewing their work as an alternative model of behavior, it was criticized and marginalized. For example, Gilligan makes it clear that in Western culture, the goal for psychic development is founded on a model of separation where the individual seeks to be able to stand alone in a kind of heroic attitude. Gilligan's research illustrates how women's tendency to focus on maintaining their relationships, sometimes at the expense of achieving some goal, has been pathologized when it is viewed using the masculine model as the norm; it suggests an inability or unwillingness to separate psychically from one's community. Understanding and accepting this theme of relationality as normal is critical for understanding the nature of the women alchemists' work and suggests why it has been dismissed.

The importance of interconnectedness and the feminine's relational aspect to others is discussed further by Joan Borysenko as she examines spiritual and psychological growth in relationship to gender issues. Focusing on the Genesis story of Jacob's ladder, Borysenko recalls that Jacob dreams of angels ascending and descending a ladder from heaven. In the dream, God appears to Jacob and decrees that the land on which he is sleeping shall belong to him and to his descendants. Furthermore, God declares that He will be with Jacob always. Upon awakening, Jacob is stunned by his experience of the *numinosum* (a profound spiritual experience) and builds an altar on the place he slept, naming it Beth-el. Borysenko writes:

> The archetype of male psychospiritual development is encapsulated in this biblical hero's journey. Jacob must leave home, die to his old self, face tremendous challenges, find new meaning, and then return to give wisdom and direction to his people. In all modern-day conceptualizations of the male life cycle, the central theme of development is the separation and autonomy that Jacob's story represents.[48]

Separation and autonomy regarding one's relationships is consistent with the theories of many male theorists such as Jean Piaget and Erik Erikson. Men tend to be more focused on how they measure up in the eyes of society whereas women tend to focus on a concern with inter-personal relationships. This idea of interconnectedness will continue to be developed as the manifestations of women's alchemy are described in this book.

Borysenko argues further that typically, women immerse themselves in their spirituality by creating community and building relationships. Women also tend to look inward and engage in introspection, rather than outward for their spiritual answers.

> From my feminine perspective, the spiritual journey is less a matter of climbing up than of looking in and discovering the Inner Light that has been there all along. And once we have discovered the Light, its dissemination is more natural than heroic, more circular and relational than an autonomous, ladderlike process.[49]

Borysenko reminds us that these so-called masculine and feminine approaches to spirituality are found in both men and women, a Jungian theory mentioned earlier in this chapter. Depth psychologists refer to the *animus*, the masculine principle in women, and the *anima* or the feminine principle in men. Most spiritual systems emphasize the importance of a masculine model, focusing on the hero's journey in their stories. Borysenko believes that the feminine model of relationality needs to be explored more deeply. It is more intuitive and less ego-driven, letting the unconscious bubble up in one's life. Relationality allows greater contact with the unconscious, thus providing greater access to the totality of the psyche. I will examine this trend in greater detail when I relate the stories of Katherine Boyle and Dorothy Moore, whose alchemy seems to have been tied to their spiritual practice as Millenarians, fervent watchers for the Christian apocalypse. The anticipation of the Christian apocalypse and the associated studies of alchemy and the *Kabbalah* present an intriguing aspect to the notion that alchemical practice manifested in myriad ways.

Thus, women's alchemy, if that term is to be embraced, and its focus on experimentation, learning about the nature of the universe and developing healing medicines for one's neighbors, is not the problem it has been made to be. Ironically, Paracelsus, a well-known male alchemist and physician is accepted quite readily in the alchemical literature, despite the fact that he practiced what is called iatrochemistry or chemistry that dealt with healing. We find in Paracelsus' 1663 edition of *His Archidoxis or, chief teachings; comprised in ten books, disclosing the genius way of making quintessences, arcanums, magisteries, elixirs, etc.* that directions for healing sit right alongside those for the transmutation of metals.[50]

There appears to be a disconnect on how alchemy has been perceived and defined by scholars. The problem is further exacerbated when women's practice of alchemy is compared to a masculine norm, created by mostly male historians. Hunter and Hutton propose that the more general definition of *scientist* needs reconsideration. They suggest that the tale of the evolution of science, passed down through generations has focused on famous men at the expense, not only of women, but also of other practitioners considered of lesser station such as craftsmen, midwives and numerous types of assistants.[51] Historically, women's studies in science have been under-represented which indicates a culture out of balance and signifies an incomplete understanding of society's psychic development. To paint a broad picture that women were not educated and rarely published their work ignores the many who did find ways to pursue their studies. In the same manner, the story of alchemy and its connection to psyche is incomplete if we ignore or dismiss a group of its practitioners—whether their intent was to further science, to cure the sick, or to produce gold.

One of my goals in this book is to flesh out that picture with what I have learned about the women I have so far encountered in my research, presenting what is documented as well as that which seems logical in my thinking. This can be a tricky exercise, prone to projections and personal biases. I will do my best, of course, to separate fact from personal reflection, but both will have a voice.

Invoking the Historical Imagination

I had been researching and writing this book on the women alchemists for nearly four years when I made my first trip to Great Britain to see and experience the places where many of these women lived and studied. Similar to what I describe in the Preface, I had been feeling stuck and unable to go forward with finishing this book that was about two thirds complete; thus, I eagerly anticipated traveling to Oxford and Cambridge to soak up what I thought would be a heady atmosphere of discovery and learning. After all, both universities have a rich tradition of alchemical practice in the work of natural philosophers such as Robert Boyle and Isaac Newton. After a brief stay in Edinburgh, I set off for Oxford, looking forward to experiencing the historical imagination described so beautifully by Ruth Meyer in her book, *Clio's Circle*. She writes:

> We historians are placed in a deeply frustrating position, continually chasing ghosts and trying to reconstruct events from shadows and fleeting glimpses of the past . . . historians need to learn to live in two worlds at once. In their work, they are required to travel back in time by means of their imagination. They must try to dialogue with the ancestors, even if the ancestral voices are forever just out of earshot, and the ancestral forms appear as mere shadows.[52]

I hoped that walking in some of the same cities and locations visited by these women, who now seemed like friends to me, would be the inspiration I needed to continue my writing with more focus and purpose. As Meyer reminds us, Clio is a Muse and they do not make our work so simple. Another writer, V. Sackville-West describes his research on his ancestor, Lady Anne Clifford.

> Few tasks of the historian or biographer can be more misleading than the reconstruction of a forgotten character from the desultory evidence at his disposal, yet into no task does he rush so glibly or with so much assurance. We should ourselves be sorry to think that posterity should judge us by a patchwork of our letters, preserved by chance, independent of their context, written perhaps in a fit of despondency or irritation, divorced, above all, from the myriad little strands which colour and compose our peculiar existence, and which in their multiplicity, their variety and their triviality, are vivid to ourselves alone, incommunicable even to those nearest to us, sharing our daily life. We should read, were it permitted us to return for an hour to earth and to find the survival of our fame enshrined in some small volume or embalmed in the notebook of some devout descendant, we should read with amazement the unrecognizable rendering, the piously and unconsciously garbled version. 'What!' we should say, 'was I indeed so intractable, or so vacillating, or so deserving of pity?' . . . . Add to this, that the biographer is dealing with an age not his own; steep himself as he may in the atmosphere of his subject, be he never so detailed and accurate, he will still be in the position of a man trying to impart a craft he has never practised or to describe a country he has never seen. . . .Still, since within our limitations, it is necessary to arrive at some conclusions, certain facts do emerge, which, hanging together with as much consistency as may reasonably be asked of the shilly-shally of human nature, enable us to build up a portrait of perhaps sufficient resemblance to the original, always remembering that a portrait, even when painted from the life, is no more than an interpretation according to the individual conception of the painter.[53]

I arrived in Oxford and began wandering the ancient streets that are utterly enchanting for one who has admired the colleges and old streets shown in film. Yet, I also began to feel disgruntled and out of sorts with the old colleges that seemed to hold answers that were unavailable to me. Many of the colleges were closed to visitors due to final examinations but I was astonished that these venerable institutions, to which I had attached a kind of numinous quality, stooped to charging visitors to stroll the grounds. "What is this?" I thought. "These are wealthy, private schools, and they are charging me to see inside?" I was put off and my complexes emerged in full force. I grew up in a very modest, middle class home where private school was not even an option. How could these *wealthy people* need my £ 5, which, translated to ten American dollars at the time. However, I paid my entrance fee as the need to enter was greater than my ire.

I loved the old buildings and the campuses are beautiful but my complexes were further stirred as I was informed that many areas of picturesque grass were only open for walking by Fellows of the College. I became obsessed with the grass restrictions and wished to ask a Porter, "Don't you sometimes just want to roll around on the grass when no one is looking?" I wrote in my travel journal, "It's nutty because despite having two Ph.Ds, I feel out of place and very intimidated by this ancient place of learning. I'm convinced that even the undergrads are more educated and smarter than I!"

I continued my exploration of Oxford, exhilarated with the town and at the same time, battling my complexes around money and privilege. While having tea, I overheard some students having a conversation that was disappointing to me as this seemingly heady discussion also included some extremely offensive, racist rhetoric. I began to feel very uncomfortable with my own research as many of the women alchemists I was studying were often from the upper social classes and would be considered extremely privileged. I began to question why I wanted to write about women who very likely would have thought me unsuitable for the friendship I had imagined. How would I finish a book about people who I might find I did not even like? This was not a good omen for my writing. Yet, my work in depth psychology, a field that will be addressed more fully in a later chapter, has taught me to hold conflicting views in balance and to let matters cook awhile before making a decision.

I continued to visit the places I had read about and longed to know. I felt no closer to the women alchemists, however, and in fact, felt a little abandoned by them and deflated. I did not find them at the History of Science Museum and neither were they walking the cloisters of Christ Church. Taking some time to sit in the grass by the River Thames where the Oxford rowers were working hard in the warm sun, I finally understood a part of my question. I wrote an email to my family where I described my epiphany.

June 7, 2008: I headed to the Museum of the History of Science . . . . There are tons of telescopes, astrolabes, chemistry, and physics paraphernalia and even

> *some alchemy items but not much. I enjoyed running across a painting of the*
> *astronomer, Tycho Brahe, but his sister, whom he loved dearly, and who*
> *worked with him a lot, was not even mentioned. Typical. It is strange, but I had*
> *kind of hoped that wandering the streets of Oxford would put me in touch with*
> *the women alchemists but then realized that this wasn't going to happen as they*
> *weren't allowed to attend school here! I think I found them walking along the*
> *river, reading under a tree, while sitting on the soft green grass.*

This grass was not on a campus and the public was not excluded from wandering among the trees and meadows by the river. Anyone could sit and think
thoughts, both great and small.

Later, a colleague pointed out that my experience echoed that recorded by
Virginia Woolf in *A Room of One's Own* where she relates a story about thinking and not paying attention when she walked across the forbidden grass in an
Oxbridge (her merging of Oxford and Cambridge) college. The porter chased
her off and she writes:

> His face expressed horror and indignation. Instinct rather than reason came to
> my help; he was a Beadle; I was a woman. This was the turf; there was the
> path. Only the Fellows and Scholars are allowed here; the gravel is the place
> for me.[54]

She was also chased out of the library. Fortunately, some things have changed
and I was allowed into the Wren Library without an escort. In fact, it was one of
the Porters who encouraged me to visit.

My trip taught me much about myself but the contribution to working out
the why and what of my research was profound. I thought back on *Clio's Circle*
and knew right off that Ruth Meyer was speaking about the kind of historical
writing I wanted to do. I am well trained in traditional, linear research having
completed a Ph.D. in Education at the very fine Claremont Graduate University
in Claremont, California. My dissertation was a qualitative research study in
which I interviewed teachers about their practice; it was straightforward compared to the dissertation I completed for my Ph.D. in Depth Psychology. For that
work I explored Baruch Spinoza, alchemy, and the unconscious, certainly topics
that were anything but linear. I felt that in order to find women alchemists, I
needed to move into less traditional territory but did not know how. On my trip
to the United Kingdom I visited universities, museums and libraries, institutions
that in the past had always been my places of refuge. When confronted with
questions and challenges, I had been able to count on them for my answers.
Since childhood, books have held a numinous importance for me and it was very
disconcerting to be confronted with needing something that my old friends were
not able to supply.

Meyer references the mythological work of depth psychologist, Ginette
Paris, who writes about Apollo, explaining that he interacts with the seeker in a
linear and straightforward manner. Hermes, on the other hand, can roam and

wander in myriad ways before he gets to his final meaning. Meyer states, "Hermes is characterized by paradox, ambiguity, and fleeting insights. His winged sandals guide us on our twisting journeys to the past."[55] I realized that a challenge with my writing was one of not recognizing the conflicting qualities in my work. I was trying to create a dance between Hermes and Apollo. Apollo is business; Hermes is serendipity. How could I assist their interaction? I did not want to leave either of these energies out of my work. I could articulate that I wanted the women's stories to include the context of their lives, their *Apollonian* information. The meaning of it all would come from Hermes; however, where was the feminine voice in the midst of these two gods? I wondered if Clio could be a mediator between the linear and non-rational or did she need to act as a different voice entirely? I was not even certain what I meant by all this rumination. It was clearly time to keep cooking and stirring the pot, allowing the elements to simmer.

During the same period of time, my attempt to find a publisher for the women alchemists was likewise frustrating. My work was judged either too academic or too depth psychological. I presented a few papers on my work at a conference for historians of science that were received with some interest until I mentioned my Jungian background. It seems a majority of science historians dislike, and I would argue, misunderstand, Jung's analyses of alchemy. I could not seem to bridge the gap, which is ironic considering Meyer's note that "in 1957 . . . William Langer, the President of the American Historical Association, urged that the 'next assignment' for historians was to use the insight of depth psychology in their work." [56] I felt I had gone a long way in achieving that goal.

My work in depth psychology, an area of study that explores the unconscious part of our psyche, began to emerge as a new voice to which I had not been paying enough attention. I needed to carry out my research and writing off the linear path. It was time to move into the labyrinth where progress would not always be straightforward but fraught with twists, turns, and unexpected but deeper insights. That was where I would find my way, either into the work or possibly, out of it. Perhaps the women alchemists and I would not be going any further. Yet, I knew these women's struggles had animated me in the beginning of my search for their stories. I thought about Margaret Cavendish, sitting through demonstrations by the distinguished Fellows of the Royal Society in London, only to have her dress be the focus of their criticism. Sophie Brahe's devoted admiration of her older brother, Tycho, mirrored my own. Speaking from my feminist position, I wanted to write about their work that had been dismissed and ignored. I hoped that my book would add vital data to the alchemical literature that scarcely mentions the work of women. Furthermore, my book might inspire young girls to study science. That was where I had first begun my work but I had not seemed able to articulate that I also wanted to understand *why* these women studied alchemy. What did their work mean to them? What was their story? That was the dance between Apollo and Hermes but working with Clio would be the key for me to partner up with those conflicting energies.

Virginia Woolf describes her search for the answer to her question regarding why there seemed to be a dearth of women writers. How could it be that she could not find poetry or drama written *by* women who could be read alongside their brothers, such as Shakespeare in the sixteenth century? It troubled her as she observes that there were such strong female characters in literature going back to antiquity. "She pervades poetry from cover to cover; she is all but absent from history."[57] Woolf pondered why she could find nothing about women's history prior to the eighteenth century and we moderns can remember that after all, it was only 1929 when she wrote about this topic. But as Woolf wonders, "the life of the average Elizabethan woman must be scattered about somewhere, could one collect it and make a book of it."[58] Woolf concluded that women would not have been allowed to publish or even be creative in public. She creates a fictional sister for Shakespeare, Judith, who is brilliant and driven to act and write. Judith ends up pregnant by the theatre manager and kills herself.

Yet, Woolf believes there must have been women who were creative and brilliant. She concludes: "When, however, one reads of a witch being ducked, of a woman possessed by devils, of a wise woman selling herbs, or even of a very remarkable man who had a mother, then I think we are on the track of a lost novelist, a suppressed poet."[59] She also speculates that many authors who signed off as Anonymous were women. Woolf argues that men might experience indifference to their work but women who stepped outside society's norms were met with enmity. "The world did not say to her as it said to them, Write if you choose; it makes no difference to me. The world said with a guffaw, Write? What's the good or your writing?"[60] In a later chapter, you will see how Margaret Cavendish exemplified the vitriol a woman writer could experience!

It became clear to me that of course I would not find the women alchemists in the universities that they were not allowed to attend. A few of them were found in the library if they wrote recipe books that recorded their alchemical work and someone had gone to the trouble of saving them. Sometimes their voices emerged in stories people had written of their relationships. A few were actually acknowledged as practicing alchemists. That is how I will present their work in the subsequent chapters. I will begin with the story that caught my interest and weave contextual information in as needed. It should be clear why I have loved learning about these women and have come to have a great regard for their struggle to fit into a society that was so limiting for them. I hope that by the end of this book, the reader will appreciate why I finally understood that perhaps in their time, the women alchemists might have been socially restricted and unable to work with someone like me. However, in the context of this book, they are now in my time where I can appreciate their quirks, phobias, and failings and see that they might even be grateful to be befriended by me.

Moving Forward

The nature of alchemy will be examined in detail in the following chapters; however, for most readers, the term evokes the image of turning lead into gold. This was accomplished by creating the Philosopher's Stone that acted as the catalyst for the gold making as well as the creation of the Elixir of Life, a remedy that was thought to be able to extend life indefinitely. The emphasis on physical alchemy, or what Robert Multhauf labels "inorganic alchemy"[61] in which the goal of the work is to transmute base metals into gold, grounds us in the realm of the history of chemistry, with alchemy acting solely as its unsophisticated precursor. Many historians still refer to alchemy as a pseudo-science. Examining alchemy from a depth psychological perspective, Carl Jung points out that there are also psychological aspects of alchemy that reside in the realm of theology and philosophy, often called spiritual alchemy. A split between practitioners of physical and spiritual alchemy that will be discussed in a later chapter, paved the way for the development of chemistry as a science in the modern sense. All the same, dismissing the role played by the spiritual aspects of alchemical practice ignores half the story. It is clear that organic and inorganic alchemy (spiritual /physical) were the *scientific* or philosophical paradigm for explaining the nature of the universe well into the eighteenth century. One pondered reality within an alchemical context, much the way we presently use our modern theories of physics as a basis for comprehending our universe. Thus, just as we understand the way current research in physics finds its way into everyday thinking about existence, we can be enlightened by considering how the rules of alchemy dictated thinking about the universe in earlier times. And, finally, we can begin to see how our present notion of reality is not so disconnected from the past.

The history of the sisterhood of the *soror mystica* spans centuries, reaching backward and forward in time, merging with the history of women's contribution to Western science. At the hands of the established, Western patriarchy, many of these women endured physical suffering, usurpation of their work, humiliation, and death. Hypatia, for example, was a mathematician, astronomer, and philosopher. She lived in Alexandria in the 4th century and became a target for a religious fanatic. For her work, Hypatia was drawn and quartered—in a church. Anna Maria Zieglerin (c. 1550-1575) met a similar fate as Hypatia, being strapped to an iron chair and burned alive as a result of her failure to create the Philosopher's Stone. Yet, some women did attain a degree of scholarly respect. A young Jewish woman of Alexandria (1st century CE) known as Maria Prophitissa, or sometimes as Maria Hebraea, conducted her alchemical research unharmed and was accorded great acclaim by her alchemical brothers. Furthermore, a group of lesser-known women strode through the scientific and philosophical developments of the sixteenth through eighteenth centuries right alongside more familiar names such as Tycho Brahe, Isaac Newton, Robert Boyle,

Gottfried Wilhelm Leibniz, François-Marie Arouet /Voltaire and Johann Wolfgang von Goethe.

These women alchemists, as well as many others, sought to make sense of
the nature of their world, their theology, and their science. They seemed to have
found an outlet for their tremendous intellect and their exploration of profound,
spiritual matters; yet, they have received scant attention, even outright denial
that they studied their craft and that it was *true* alchemy. I hope to make a space
for their voices in this book: to tell their stories, to examine their reasons for
studying alchemy, and finally, to speak about their work in the context of their
feminine experience of natural philosophy—the old term for what we now refer
to as science.

A Note about Names

It is a common practice in academic writing to refer to people by either their full
name or their surname. Thus, we have Sophie Brahe or Brahe, not Sophie. Lady
Ranelagh can be Ranelagh or Katherine Jones but not Katherine. This practice
has been the norm for writing about men which can be justified, as usually, their
surname is also their birth name. Isaac Newton can be Newton and his identity is
intact, although some might argue that even the surname distances one from
knowing the total person. In the course of my writing, these women have become more than just a name and I refer to them in that manner; thus, Dorothy
Moore becomes Dorothy, not Moore. Furthermore, many of these women were
less than fond of their husbands and distilling their persona to a surname that is
not a birth name does them a disservice. A major point of this book is to provide
a space for these women's voices and that entails honoring them in a way that is
contrary to the patriarchal tenor of academia.

Notes

1. Marie Meurdrac, *La Chymie charitable et facile, en faveur des Dames Charitable, 1666* [*Easy Chemistry for Women*] (Paris: CNRS Editions, 1999), 17.

2. Reinhard Federmann, *The Royal Art of Alchemy*, trans. R. H.Weber (Philadelphia:
Chilton, 1964/1969).

3. Michael White, *Isaac Newton: The Last Sorcerer* (Reading, MA: Perseus Books,
1997).

4. Reinhard Federmann, *The Royal Art of Alchemy,* trans. R. H.Weber (Philadelphia:
Chilton, 1964/1969).

5. Margaret Alic, *Hypatia's Heritage: a History of Women in Science from Antiquity
Though the Nineteenth Century* (Boston: Beacon, 1986).

6. Marelene Rayner-Canham and Geoffrey Rayner-Canham, *Women in Chemistry:
Their Changing Roles from Alchemical Times to the Mid-Twentieth Century* (Philadelphia: American Chemical Society and the Chemical Heritage Foundation, 1998).

7. Alic, 20-21.

8. Alic, 22.

9. Ali, 22.

10. Alic, 25-26.

11. Stanley Rubin, *Medieval English Medicine* (London: David and Charles, 1974).

12. Rubin, 186.

13. Rubin, 186.

14. Rubin, 187.

15. Rubin, 187.

16. Robin L. Gordon, "The Murder of Spinoza and Other 17th Century Alchemists: A Contemporary Look at a Long-Ago Mortificatio Tale" (PhD dissertation, Pacifica Graduate Institute, 2004).

17. *Separatio* refers to the separation of the different types of matter in the alchemical vessel or what we would call today, the beaker.

18. Carl G. Jung, *Psychology and Alchemy,* 2nd ed. (Princeton, NJ: Princeton University Press, 1967).

19. Raphael Patai, *The Jewish Alchemists: A History and Source Book* (Princeton, NJ: Princeton University Press, 1994) and Margaret Alic, *Hypatia's Heritage: A History of Women in Science from Antiquity Though the Nineteenth Century* (Boston: Beacon, 1986).

20. Allister McLean, *A Commentary on the Mutus Liber* (Grand Rapids, MI: Phanes Press, 1991), 7.

21. Carl G. Jung, *Psychology and Alchemy*, 2nd ed. (Princeton, NJ: Princeton University Press,1953/1968); Patai (1994).

22. Lucia Tosi, "Marie Meurdrac: Paracelsian Chemist and Feminist," *Ambix* 48 (July 2001): 69-82.

23. Margaret Alic,1986; Lynette Hunter and Sarah Hutton, 1997; Weisner, 2000, and Tara Nummedal, 2001.

24. Clark Moustakas, *Heuristic Research : Design, Methodology, and Applications* (Thousand Oaks, CA: Sage, 1990).

25. Clark Moustakas, *Phenomenological Research Methods* (Thousand Oaks, CA: Sage, 1994).

26. Moustakas, 17.

27. Edward F. Edinger, *Archetype of the Apocalypse: Divine Vengeance, Terrorism, and the End of the World* (Chicago: Open Court, 1999), 135.

28. See Edward F. Edinger, *Anatomy of the Psyche: Alchemical Symbolism in Psychotherapy* (Chicago: Open Court, 1985); Marie Louise Von Franz. *Alchemy* (Toronto, Inner City Books, 1980); James Hillman, *Re-visioning Psychology* (New York: Harper Collins, 1975).

29. *Coniunctio* is the alchemical step in which the purified elements, both dry and liquid, are reunited to eventually create the Philosopher's Stone.

30. Lynette Hunter, "Sisters of the Royal Society: The Circle of Katherine Jones, Lady Ranelagh," in *Women, Science and Medicine 1500 - 1700: Mothers and Sisters of the Royal Society*, Lynette Hunter and Sarah Hutton, eds. (Gloucestershire: Sutton, 1997), 178-197.

31. Robert P. Multhauf, *The Origins of Chemistry* (New York: Franklin Watts, 1966).

32. Alic, 36.

33. Lynn L. Thorndike, *A History of Magic and Experimental Science, Volumes VII and VIII: The Seventeenth Century* (New York: Columbia University Press, 1958), 242.

34. W.S.C. Copeman, *Doctors and Disease in Tudor Times* (London: Dawson of Pall Mall, 1960), 108.

35. Copeman, 110.

36. Copeman, 110.

37. Peter J. French, *John Dee: The World of an Elizabethan Magus* (London: Routledge and Kegan Paul, 1972), 127.

38. Tara E. Nummedal, "Alchemical Reproduction and the Career of Anna Maria Zieglerin, *Ambix*, 48 (July 2001): 57.

39. Patricia Crawford, "Women's Published Writing 1600 – 1700." In *Women in English Society 1500–1800,* Mary Prior, ed. (London: Methuen, 1985), 211 – 264.

40. Sara Heller Mendelson, "Stuart Women's Diaries and Occasional Memoirs." In *Women in English Society 1500 – 1800,* Mary Prior, ed. (London: Methuen, 1985), 181 - 210.

41. William Bray, ed., *Memoirs of John Evelyn,* Volumes 1-4 (London: Henry Colburn,1827).

42. Crawford (1985).

43. William Leibniz's work, *Opuscula Philosophica* (1690) is more accurately based on Anne Conway's work. Van Helmont was given credit for Conway's manuscript, *The Principles of the Most Ancient and Modern Philosophy* (1692).

44. Hunter and Hutton, 1997.

45. Lynette Hunter and Sara Hutton (1997).

46. Alan Cook, F.R.S. "Ladies in the Scientific Revolution." *Notes and Records of the Royal Society,* 51 (January 1997): 8.

47. Carol Gilligan, *In a Different Voice* (Cambridge: Harvard University Press, 1982/1993).

48. Joan Borysenko, *A Woman's Journey to God* (New York: Riverhead Books, 1999), 71.

49. Borysenko, 71.

50. Paracelsus' 1663 edition of *His Archidoxis or, chief teachings; comprised in ten books, disclosing the genius way of making quintessences, arcanums, magisteries, elixirs, etc.englished by J.H. Oxen* (London: Lodowick Lloyd, 1663), 31- 76.

51. Hunter and Hutton (1997).

52. Ruth Meyer, *Clio's Circle* (New Orleans: Spring Journal, 2007), 22-23.

53. V. Sackvile-West, *The Diary of the Lady Anne Clifford with an Introductory Note by V. Sackville-West* (London: William Heinemann, 1923), xxiv-xxvi.

54 Virginia Woolf, *A Room of One's Own* (New York: Harcourt,1929/1957), 6.

55. Meyer, 10.

56. Meyer, 11.

57 Woolf, 43.

58 Woolf, 45.

59 Woolf, 49.

60 Woolf, 52.

61. Multhauf,12.

Chapter 2

Alchemy 101: Connecting an Ancient Tradition to Modern Thinking

> *The name of Philosopher has been given from time immemorial to all those who are duly initiated into the operations of the Great Work, which they term also Hermetic Science or Philosophy, because Hermes Trismegistus is regarded as the first who became illustrious therein.*[1]

The question I raised in the first chapter, why would anyone in the 21st century care to understand a science so ancient as alchemy, can be best addressed with a modest foundation of knowledge about alchemy. In this chapter I will explain how the philosophical foundations for alchemy are still quite relevant and active in modern society. However, we first need to begin by establishing some background regarding what alchemy was and is. I will discuss a few well-known male alchemists in order to clarify why we need to think of alchemy as a legitimate endeavor rather than with the typical label, *pseudo-science*. The relationship of alchemy to the spirit-matter split that emerged in the seventeenth century will also be examined. Finally, I will examine what depth psychologists such as Carl Jung, Edward Edinger, and Marie Louise von Franz concluded about the psychological aspects of not only alchemy, but of scientific endeavor in general. This last section brings alchemy into our modern era and explains why alchemical principles can be useful in thinking about the 21st century.

Attempting to define alchemy confronts the reader with a motley blend of descriptions as to its nature. Alchemy holds the distinction of being the precursor to modern chemistry. However, a more familiar image may be that which comes to mind of a wizard-like fellow, standing over his retort (vessel used in alchemy), wondering if *this* time he will succeed in transmuting the lead into gold. Alchemy was the process used to create the Philosopher's Stone and the Elixir of Life. The Philosopher's Stone was critical in the transmutation process. It was considered that just a morsel the size of a grain of sand could turn base metals such as lead into gold or if added in the final step in the tincture making process, create a healing elixir that could extend life indefinitely. A catalyst is the term we use in modern chemistry for the role of the Philosopher's Stone. The processes for creating gold or the elixir are very similar; however, the latter uses vegetable matter as opposed to minerals or metals as the substance to be worked upon.

Alchemy has been portrayed as quackery, charlatanism, and dark magic. Secret societies have used alchemical symbolism in their rituals, practices, and mythologies. Carl Jung saw in alchemy, an analogy for the individuation process in psychological development and concluded that alchemists projected something of their unconscious into the understanding of their art. Many alchemists believed that their art was revealed to man by God and thus, only an alchemist who was worthy would be graced by God to achieve the highest result, the Philosopher's Stone. It was not to be mistaken for magic, although many an alchemist became seduced by unrelated magical practices, probably adding to the misconception of associating alchemy with witchcraft. Heinrich Khunrath, a sixteenth century alchemist, "interpreted transmutation itself as a mystical process occurring within the adept's soul."[2] Khunrath includes a fascinating depiction of an alchemist in *Amphiteatrum sapientiae* (1609) in which the Adept (the term often given to alchemical practitioners) is seen kneeling in front of an altar, arms outstretched and apparently asking a blessing on his work. Henry Cornelius Agrippa wrote about the issue of the magus in his *Fourth Book of Occult Philosophy* (1655) in which he made it a point to distinguish between the black arts and alchemy. He argued that the black arts required the practitioner to be in league with Satan. Agrippa states that the word magus was Persian and referred to a person who studied divine or godly things. He also quotes Plato as having said, "Magick is the art of worshiping God."[3]

Numerous historians have taken each of these aspects of alchemy and explored them deeply. The task can be monumental, as the alchemical tradition can be traced back thousands of years to before the time of Socrates and probably is as old as humankind's propensity to experiment with herbs in brewing remedies to heal sickness. Although the history of practitioners of ancient alchemy provides interesting reading, the goal of this chapter is to provide a basic understanding of the nature of alchemy in order to give the reader enough background information to be able to make sense of this topic that so intrigued the women being discussed in this book.

Western alchemy traces its roots to ancient Greece, Rome, and especially Egypt; however, China and India have their own rich alchemical traditions.[4] Michael White claims that the earliest records of creating an Elixir of Life came from China.[5] 5. Masumi Chikashige explains that Chinese alchemy began with the goal of finding the elixir of life and creating gold was secondary. The gold production came out of the idea that it was thought to be an essential ingredient in the elixir.[6]

Robert Multhauf explores the origins of chemistry, describing a timeline of development marked by various achievements in metallurgy, glass making, pottery, and practical chemistry or alchemy.[7] Traditionally, alchemy is associated with Thoth (Egyptian god) also known as Hermes. Alchemy was not known by that name until much later but the art of transmutation is discussed in a number of ancient manuscripts attributed to writers such as Zosimos, Democritus and Avicenna (Persian, c.1307). Al-Farabi, a Turk (c. 950), is linked with writing *On the Origins of the Sciences* where Multhauf notes, "He lists as one of eight parts of physics, 'alchemy, which is the science of the conversion of things into other species.'"[8] Jãbir ibn Haiyãn who lived around 700 CE is credited with translating Arabic alchemy into Latin providing access to the art by Western alchemists.

Regardless of the country of origin, Raphael Patai explains that alchemy "was built on the theory that all the visible forms of matter, whether mineral, vegetable, animal, or human, were manifold forms of one basic, essential substance."[9] This substance was called the *prima materia* or first matter. Countless descriptions of the *prima materia* sound as if it is some type of cosmic *stuff* that exists and is associated with God.[10] The first matter manifests itself in some mysterious way to become our physical world. Marie Louise Von Franz writes, "that if this basic material could be discovered, one would, in a way, look into the divine fabric of the cosmos."[11] Richard Federmann adds, "Gold according to alchemic theory is contained in every other metal, and each is but a visible form of the Aristotelian prime matter."[12] This search for our species' origins will take on added significance when discussed in the context of the Scientific Revolution.

Alchemy was based on the belief that matter was made up of a combination of three components, sulphur (soul), mercury (spirit or life force), and salt (physical body). In this process, sulphur and mercury are not synonymous with the actual elements. They represent aspects of matter that carry a kind of fire or spirit. Reinhard Federmann asks:

> Why did they choose sulfur and mercury? Was it because these were the elements the early alchemists used to heat in their retorts? Sulfur sublimated [vaporized], yet remained sulfur; fire had no effect on it; it had an affinity with fire and therefore also with the sun and thus with gold. Mercury evaporated, yet returned to the liquid state; it was a metal. Like water, mercury had an affinity with water and thus with the moon that rules over water (the oceans); it had the color of silver, just as sulfur was yellow as gold. To combine sulfur with mercury became the next step – unite them in certain proportions; marry the two, man and wife, sun and moon, gold and silver, fire and water; and the result must be the quintessence![13]

The alchemical operations consisted of separating a substance into the three basic components before recombining them to create a new substance that held the three original elements in perfect balance or harmony. This substance was often called the *quinta essentia* or *quintessence* and was thought to be analogous to the Philosopher's Stone.

In Figure 2.1 the small triangles represent sulphur, mercury and salt which are enclosed in a circle that is the impure substance to be worked upon. They are separated into their pure essence and then recombined to create a pure form of the substance. It was thought that all matter, including humans, could be made pure (i.e., healed from the impurities that made it sick).

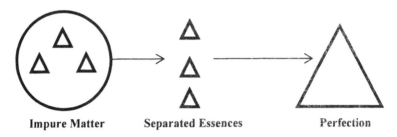

| Impure Matter | Separated Essences | Perfection |

Figure 2.1: Figurative Representation of the Alchemical Process

For example, in the same way that plants and animals grow from embryos, gold (believed to be a pure substance) was thought to begin its journey as a less perfect metal that gradually developed into the pure form that is coveted by so many cultures. Gold was considered the most pure metal since it has unique properties. For example, gold is resistant to changes such as tarnishing and besides, was found in the ground in its natural state.

Alchemy was also tied directly to astrology in that the planets were believed to rule all life such as the formation of plants and minerals. As the planets made their transits through the Zodiac, life on earth would be affected in diverse ways according to the planet's characteristics. The movement of planets could portend peace, war, and prosperity to name a few. Consequently, each planet became associated with an Earth element and herb. The sun was associated with gold, a rather obvious connection. Lavender was associated with the planet Mercury. It was believed that astrology indicated the best time to collect the specimens for experimentation. The association with astrology goes back to alchemy's inception. The astrological signs for the planets are used in alchemical shorthand to this day.

Summary of the Alchemical Operations
In Contemporary Language

A brief summary of creating a medical tincture will serve to illustrate the alchemical operations. There are numerous sources for the steps but the following have been condensed from directions given by Frater Albertus in *The Alchemist's Handbook*.[14] I have chosen this set of directions because they are modern and explain the alchemical operations in a way that the contemporary reader can understand easily. Reading directions from the early alchemists presents a challenge to the modern reader but once one understands the basic principles of the work, one can pick out meaning in the obscure language much more easily.

Albertus tells the reader that alchemy is really quite simple in theory but the practice takes years of study and dedication. One needs a heat source, flasks, and distillation equipment which are easily purchased. The location of the laboratory is up to the individual as long as a continuous heat source is available. The attitude of the alchemist or Sage is critical. The work is approached with humility and respect. Reading directions in a book is not enough; the spirit of the alchemist must be in a harmonious state and ready for the work. The beginner should not attempt to create the Philosopher's Stone, a feat that most will never achieve and requires much preparatory practice. The beginner should also start with the plant kingdom since working with minerals and metals can be highly dangerous when they are heated and distilled.

One of the first steps in the alchemical process requires one to dissolve the matter being worked upon in some liquid. The nature of the solvent used in alchemy was critical to its success. Substances such as urine were used but dew is also often cited as a good choice. Arthur E. Waite describes the wonders of dew in the esoteric language found so often in alchemical writing.

> Of all natural bodies, dew was deemed the most powerful dissolvent of gold; and the cross, in chemical language, was equivalent to light; because the figure of a cross exhibits at the same time the three letters of which the word lux, or light, is compounded. Now, lux is called...the seed or menstruum of the red dragon, or, in other words, that gross and corporeal light, when properly digested and modified, produces gold. Hence it follows, if this etymology be admitted, that a Rosycrucian philosopher is one who by the intervention and assistance of the dew, seeks for light, or, in other words, the substance called the Philosopher's Stone.[15]

Whichever herb is needed should be gathered in full bloom and dried in a protected spot. After the plant matter has dried thoroughly, one way to extract the essential elements is to grind the dried material and place it in a container of very pure alcohol (sometimes distilled from wine or gin). The solvent is called the *menstruum* and the stronger alcohol is thought to be able to extract the maximum essential elements or the essence from the plant matter. The ancient alchemists experimented with many solvents. The goal is to separate the herb into its essential oils that carry the Sulphur principal and are soluble in alcohol. The container is kept in a warm, dark place soaking the plant matter (maceration) in the solvent for many days.

The next step requires one to filter the liquid retaining the plant matter that is still an essential part of the work. The liquid is also reserved as it contains the essence of the plant. The alchemists referred to this at *Caput Mortuum* or Dead Head, also referred to as feces. The *Caput Mortuum* is burned to ash in the operation called *calcinatio*. The ash is continually heated until it turns to a light gray or white color. It is then ground to a very fine texture. This substance is thought to be unique to the plant and is a critical component in the final stage. It is also referred to as the salt or body.

The ash is placed in another flask so that the liquid containing the essence can be added to it. The liquid that contains the essence is thought to be similar among all plants since the logic is that the essence contains the quintessence or life force of all plants. The alchemists referred to this operation as *solutio*.

The flask is sealed tightly and is heated at a low temperature for a few weeks to allow the mixture to digest. The digestion stage is referred to as *mortificatio* whereas the recombination of essence and salt is called *coniunctio*. It is believed that the salt/body absorbs the essence that is in a pure form. The resulting fluid can be used for medicinal purposes if it is diluted. Albertus concludes, "This is the most primitive and simplest form of preparing an herbal substance according to the precepts of Alchemy."[16] I have greatly simplified the work and different operations can be referred to with the same name depending upon the writer.

The preceding steps might be carried out on the same substance repeatedly in order to achieve the purest and strongest concentration of the element. Similar steps were followed using metals and minerals but the process was long and arduous. Sometimes it was explosive as can be seen in engravings or in paintings such as Heindrick Heerschop's (c. 1627) *The Alchemist's Experiment Takes Fire*.

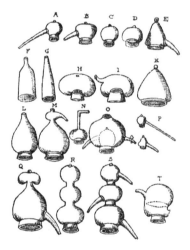

Figure 2.2: Examples of Early Alchemical Equipment

The apparatus that was developed for alchemical practice has a rich history. It is easy to take for granted the array of flasks and beakers that are available for the contemporary alchemist. Producing a substance that would seal a flask without melting in the fire was a significant step forward. Making glass containers that would not break under high heat was another breakthrough. The list is extensive. Michael White writes concerning alchemical apparatus:

Between the founding of Alexandria and the decimation of its famous library towards the end of the fourth century AD, some of the basic techniques of alchemy were first devised and popularised in texts credited to Democritus, Isis, Moses, Hermes and many others. These techniques included distillation, which is still used extensively today and stimulated the development of many common pieces of equipment such as the ambix or *alembic* (a glass or copper vessel), the *solem* (a distillation tube) and the *tribikos* (a funnel).[17]

Figure 2.3: Pelican—the apparatus and the old symbolic reference to the process.

There is an extensive lexicon of alchemical terminology that describes the alchemical process in terms that have confused readers, as was probably the intention. True alchemists understood the language whereas it was hoped that charlatans would not. Yet, once we begin to immerse ourselves in the obfuscated directions for achieving transmutation, a certain logic emerges. We may read about the King and Queen being united in marriage, and come to see it as an obvious symbol for the combination of elements.

Describing A Few Early Alchemists

The reader will notice that the following notes focus on the work of male alchemists who were far more published throughout history. Some of the earliest alchemical writings are attributed to Hermes Trismegistus, thought to have lived in Egypt around the same time Moses was alive. Many stories surround Hermes and his work but it is believed that Alexander the Great discovered an emerald tablet in Hermes' tomb on which 13 sentences regarding the *Magnum Opus* written in Phoenician, were inscribed. Line two of the Emerald Tablet is interesting in that it has become a famous alchemical saying: "What is below is like that which is above, and what is above is like that which is below, to accomplish the miracles of one thing."[18] The aphorism is believed to have meant that the physical world works the same way as that which occurs in the heavens or in God's domain. Thus, understanding the micro world would allow the practitioner to understand the universe and ultimately, God. The Emerald Tablet also explains directions for completing the alchemical work, although they are couched in the cryptic language found in older alchemical writings. The term "hermetic seal," still used today, originates from the term, "Hermes, his seal."[19]

Aristotle (384-322 BCE), Zosimos of Panopolis (5th century), various Arab adepts and numerous other practitioners contributed to the foundation of Western alchemy. Aristotle taught that matter was created from the four elements: earth, air, fire, and water. Different matter contained different proportions of the elements and could be transmuted from one to another by changing proportions of the element. These basic building blocks gave matter its characteristics of cold, dry, hot, or wet. The theory supports the idea that a metal such as lead could be turned into gold by a rearrangement of its basic structure.

Zosimos wrote profusely about alchemy, providing early records of practical information such as the description of laboratory equipment. It is from Zosimos that we learn of the prolific Jewish tradition in alchemy, naming specific practitioners such as Maria Hebraea (the woman depicted in the painting on the cover of this book) or Maria the Jewess.[20] Maria is credited with inventing or at least making famous many kinds of laboratory apparatus such as a water-bath called the *Bain Marie* and the *tribikos*, a distillation earthenware vessel. She described directions in detail for completing the alchemical operations. Her famous axiom states, "'One becomes two, two becomes three, and by means of the third and fourth achieves unity; thus two are but one.'"[21] This axiom was believed by alchemists to hold the key to successfully achieving transmutation. Jung also analyzed the axiom's psychology, but its precise meaning remains an alchemical mystery.

Jábiríbn-Hayyán, also known by Europeans who could not pronounce his name correctly as Geber, was one of the more famous Arabian alchemists. He is thought to have been born around 702 CE. He formulated the notion that:

> Metals were but different mixtures of sulphur and mercury, the precious metals being richer in mercury than in sulphur. The transmutation of lead or copper into gold or silver meant the withdrawal of sulphur from and the addition of mercury to them.[22]

Thus, gold consisted mostly of mercury and a minor amount of sulphur. Other famous alchemists from a variety of countries include Albertus Magnus (1193-1280), Raymund Lully (1234-1316), Thomas Aquinas (1224-1274), Arnaldus de Villa Nova (b.1235), Roger Bacon (1214-1294), Nicholas Flamel (1340-1418), Michael Maier (1568-1622) and John Dee (1527-1608). Alchemists were often physicians due to the emphasis on the healing properties of the tinctures they made. The reader is encouraged to explore both the myriad secondary sources on alchemists as well as to attempt to read primary sources which becomes easier with practice.

Practitioners of the Magnum Opus

Michael Sendivogius

Michael Sendivogius wrote a treatise that will serve as an example describing the way alchemy is supposed to have worked. He was born in Poland and lived from approximately 1556 until sometime after 1636.[23] His biography is sketchy and not without hyperbole but he did contribute greatly to alchemical knowledge. His treatise, *The New Chemical Light*, was published in 1604.[24] There is some speculation that he plagiarized this work from his wife's deceased former husband, the famous Scottish alchemist, Alexander Seton.[25] In any case, it is interesting to note that for some reason, Sendivogius believed that he need- ed to disguise his authorship and did so using an anagram of his name. Waite translates "Divi Leschi genus amo"[26] to "I love the Divine Race of Leschi"[27] which is the anagram for his name, Michael Sendivogius. Sendivogius wrote that he did not want to bring undue attention to himself and after all, the knowledge he had gained was through God's grace. "The facts and deductions which I have here briefly set down are transcribed from that manual— experience, graciously bestowed upon me by the Most High."[28]   It would be unseemly to imply that he had come to this knowledge on his own.

Sendivogius devised an intriguing theory regarding the *prima materia* and the formation of matter. Briefly, he concluded that the *prima materia* was a cha- otic material that existed in the inner core of the earth and that it created a seed which took the form of a vapor that traveled to the earth's surface via pores in the earth's crust. The vapor purified the earth with which it came into contact as it made its ascent. Depending on the subsequent purity of the terrain through which it traveled, the seed or vapor could form into the purest gold or less per- fect metals such as lead and copper. The winter frosts were believed to trap the vapor in the earth's crust until spring when the matter escaped and caused the plants and trees to come into bloom. Sendivogius enlightened the reader regard- ing natural science stating that all mineral, vegetable, and animal matter, thus contained the same *origin*; however, how the seed differentiated itself into the different forms is difficult to fathom from reading his work.

It will not help the reader to attempt to make sense of Sendivogius' theory within the context of our modern scientific knowledge. We can see, though, that his theory was consistent with alchemy in that the seed to which he referred was generated from a *prima materia*-like "first substance"[29] and that the world was made manifest from some naturally occurring internal mechanism. Sendivogius also stated that he believed that the alchemists of his day were extremely fortu- nate, noting that they were much more advanced regarding the information to which they had access compared to older practitioners like Hermes Trismegistus, Raymund Lully (1234-1316) or Nicholas Flamel (1340-1418), famous alchemists in their own time. Despite having access to more modern alchemical methodology, Sendivogius notes that modern alchemists had still failed to achieve the creation of the Philosopher's Stone and wonders whether the less sophisticated ways were more effective, after all.

The New Chemical Light is significant in that Sendivogius gave very specific directions on creating the Philosopher's Stone, in both the poetic as well as obscure style of the alchemists. One excerpt states:

> Take eleven grains of our earth, by as many doses, one grain of our gold, and two grains of our silver. Here you should carefully bear in mind that common gold and silver are of no use for our purpose, as they are dead.... Remember that God alone can create; but He has permitted the Sage [another name for alchemists] to make manifest things that are hidden and concealed, according to the ministry of Nature.[30]

The directions read like a recipe book, a theme that will be picked up in chapter five of this book.

Paracelsus

Paracelsus, another famous name in alchemical studies, was considered one of the great practitioners of the Art as well as a gifted physician. Jung wrote extensively about Paracelsus and his mystical notions of alchemy.[31] The work of Paracelsus is also a nice example of the state of physical alchemy practiced in the 15th-16th centuries. Paracelsus was born in 1493 in Einsiedeln, Switzerland and died in 1541. His birth name was Theophrastus Bombast von Hohenheim. His father was a physician in whose footsteps he followed, except Paracelsus did not subscribe to the traditional medical model of the time that was based on the error-plagued work of Galen. Paracelsus was part of a group of thinkers who believed that making empirical observations about the natural world was more important than blindly accepting old scientific ideas, even though great thinkers such as Aristotle generated these ideas.[32] This was revolutionary thinking for the time. Traditionally, ideas about the natural world emerged out of reflection and discourse. Paracelsus and his followers wanted to overturn that model which created an age old archetypal rift between long held and innovative philosophies.

Paracelsus traveled extensively in Britain, Sweden, Poland, Turkey, Greece, and the Ionian islands as well as in Alexandria, where he practiced medicine and learned novel techniques (for him) from many people, including practitioners of the magical arts. Paracelsus' alchemy focused on creating medicine rather than gold. He believed that disease occurred as a result of the mercury, sulphur, and salt of the body being out of balance. He held that "the alchemist was to discover the medicines and prepare them, the physician was to examine and explain their action."[33] *Paracelsus His Archidoxis or, Chief Teaching; Compiled in Ten Books, Disclosing the Genuine Way of Making Quintessences, Arcanums,*

Magisteries, Elixirs, &c. (translated from German or as he states more quaintly, "Englished" by J.H. Oxen) contains information on the transmutation of metals as well as creating the "spirit of life."[34] Illnesses of the brain, liver and spleen are addressed alongside natural philosophy. Most of Paracelsus' books were not published until after his death. During his lifetime, he became increasingly bitter and wrote heatedly against his colleagues. Jung's analysis states:

> His [Paracelsus'] exposition of a subject is seldom systematic or even coherent; it is constantly interrupted by admonition, addressed in a subtle or coarse vein to an invisible auditor afflicted with moral deafness. Paracelsus was a little too sure that he had his enemy in front of him, and did not notice that it was lodged in his own bosom.[35]

Paracelsus was 48-years-old when he died in Salzburg in 1541. As so often happens, Paracelsus' methods and philosophy gained more respect after his death with the next generation of sixteenth and seventeenth century physicians. Nevertheless, during his lifetime he had a reputation for being socially difficult albeit technically brilliant, especially regarding the treatment of syphilis, common wounds, and the practice of surgery. One of Paracelsus' treatises that was eventually published was titled *Seven Defensiones: The Reply to Certain Calumniations of His Enemies by Theophrastus Von Hohenheim Called Paracelsus*. I will include a brief summary of the seven defenses because they provide the reader with a sense of Paracelsus, both as a personality and as a passionate healer. They also describe specific practices that may assist the modern reader in understanding what the healing arts were like at the time of his writing. It has been translated from the German. In the First Defense he speaks of a new way to practice medicine.

> If there were a man sick of a fever due to end in xii weeks, and then it would be over and done with, and if it happened that the sick man demanded medicine to drive out this fever before its appointed end, he would have two kinds of physician before him, the false and the true. The false proceeds as follows: he begins gradually and slowly to doctor him, spends much time on Syrups, on Laxatives, on purgatives and oatmeal mushes, on barley, on pumpkins, on Melons, on julep and other such rubbish, is slow and frequently administers enemas, does not know himself what he is doing, and thus drags along with time and gentle words till he comes to the term. Then he ascribes the spontaneous end to art.[36]

One can see why his colleagues were often offended. Paracelsus continues his argument in the Second Defense, declaring his belief that one of the problems with medicine is that his colleagues are suspicious by nature regarding progress and do not have an understanding of astrology, a major factor in disease.

Thus too with prescriptions, they say I write them new receipts and introduce a new procedure. As they have told me to my face: I am to use nothing strange, according to the meaning of God's tenth commandment: Thou shalt not covet strange things.[37]

The Third Defense justifies Paracelsus' use of substances that his colleagues considered poison and he makes it clear that he believes his colleagues are ignorant, referring to them as "good-for-nothing tongues,"[38] hardly the way to endear oneself to one's associates. In the Fourth Defense, Paracelsus justifies his traveling around the world to learn medicine and to practice the healing arts, arguing that he was not simply the vagabond he had been called but one who is willing to learn from everyone. He continues in this vein in the next Defense, criticizing what he views as the medical establishment's greed for money and fine living. Paracelsus goes on to insult just about everyone, stating that medicine should be practiced to heal, not to make money.

Thus too come all the lazy and profligate rascals into medicine and sell their medicine, whether it makes sense or not. Now he who can fill sacks with gold, he is praised, he is a good physician. Thus the apothecaries too and some barbers take medicine upon themselves, behave and carry on as though it were a woodcart, go into medicine against their own conscience, forget their own souls, if only they become rich, prepare house and home and all that belong in it, and dress it up![39]

The Sixth Defense is slightly conciliatory, explaining that Paracelsus' upbringing was very different from his detractors and perhaps that was the reason why he seemed so out of step compared to them.

I am not by nature subtly spun, neither is it usual in my country to attain anything by spinning silk . . . .Therefore must the coarse be judged coarse, though the same think himself utterly subtle and charming. Thus it is with me too: what I think is silk, the others call ticking and coarse cloth.[40]

The Seventh Defense concludes with an understatement, to be sure, "And so, Reader, hast thou to some extent understood me in this reply and seen well that I have attacked with all mildness."[41] Paracelsus' argument with established medical practices could be traced to his belief that man's knowledge was revealed from both God and an inner "Light of Nature."[42] The Light of Nature was given to man by God and allowed him to observe and develop ideas from his own experiences. We will see this idea of an inner light surface again with the work of Jacob Böehme. Thus, Paracelsus' belief in the reality of alchemy, as well as astrology and magic, makes sense in that they have been revealed to him via this inner light. Jung writes, "There was no form of manticism and magic that Paracelsus did not practice himself or recommend to others."[43] It is no wonder that the more conservative medical establishment had its reservations about him. It is also not surprising that someone with Paracelsus' temperament and belief in the truth of inner knowledge would clash so violently with his colleagues, who merely deferred to Divine revelation.

Michael Maier

Michael Maier (1568-1622) was another physician-alchemist.[44] He was German, joined the Rosicrucian Society and treated Rudolph II in Prague for many years. Maier's *Atlanta Fugiens* is probably his most well-known work; however he wrote copiously.

Maier is an example of an alchemist who lived in the spiritual world, which is evidenced in his writings that veer toward esotericism. For example, in one of Maier's earlier tracts, *A Subtle Allegory Concerning the Secrets of Alchemy Very Useful to Possess and Pleasant to Read*, he is searching for the Phoenix which many claimed did not exist. He suggests that perhaps the Phoenix is symbolic for the philosopher's stone. Thus, Maier explains in his book that he plans to set off on the vernal equinox to Europe which he associates with the element earth; then to America (water); Asia (air), and finally Africa (fire). The actual trip did not take place but was merely a metaphor for his alchemical work. Maier uses an interesting image suggesting that if we agree that a cannon ball would take longer than eight days to travel around the 25,000 miles of the earth, and that we also agree that the much more massive sun does so in 24 hours, one must consider how fast the sun travels in its orbit as well as the other planets in the universe. He writes:

> It would make our thoughts reel if we strove to realise the velocity with which Saturn moves round the Sun, and with which the heavens revolve round their own axis. But greater still, and far more wonderful, is the speed of human thought, which, in a moment of time, travels from one end of the heavens to the other. We may believe that the angels, as spiritual beings, move with the quickness of that which is spiritual in man, viz., thought.[45]

Maier's book continues with his allegorical travels taking him to the Americas and later to the Persian Gulf where he wends his way to the Holy Land, meeting up with the spirit of Jason of the Golden Fleece. He continues to India and to Africa noting that a year has passed since he began the journey. Maier has found some of the answers to his questions as the journey has proceeded. Finally, near the Red Sea he asks the Eryrhraean Sibyl about finding the Phoenix. She sends him to find Mercury in one of the seven mouths of the Nile. Maier is happy and sets out again but finds nothing. He reasons that he has misunderstood what he has been told. He sets out yet again and finds Mercury, who shows him the dwelling of the Phoenix.

21st Century Alchemists

It may come as no surprise to the reader that alchemy is still being carried out as both a science and a spiritual practice in the 21st century. Books such as *The Practical Handbook of Plant Alchemy* by Manfred Junius[46] and the *Alchemist's Handbook* by Frater Albertus[47] assist the reader in preparing tinctures used for healing a variety of ailments. An example from Albertus' work was used to illustrate the alchemical operations in a previous section. Alchemical web sites also provide modern, detailed directions on carrying out the Art. For example, in executing the steps for the Lesser Circulation (alchemy practiced with plants), Mark Stavish is quite detailed as he explains that one merely needs dried herb, pure grain alcohol, and an assortment of household cooking implements.[48] The herb is picked according to the medicinal need and must be harvested at the right time of year, depending on whether the leaves or flowers are needed. The work is also begun with prayer. "Start with prayers to God that the mysteries may be revealed to you and your place in the Universe restored."[49] The herbs are dried, soaked in alcohol, and subjected to a period of about two weeks of evaporation and condensation. Further operations include separating the plant matter from the fluid portion, distillation of the fluid, calcining (burning) the plant matter to ash, and recombining the fluid and ash to form the healing elixir. This is a very brief summary of the work and is similar to the directions found in old manuscripts that theoretically, if the steps were carried out properly, created the Elixir of Life that had extraordinary powers of healing.

The steps for working with minerals are basically the same as with plant matter; however, they extend beyond the first recombination phase and include more complex operations that result in the Philosopher's Stone. The work with minerals often involved using mercury, lead, and antimony which can be poisonous when volatile. Although the alchemical Art may have been updated, the modern writers maintain the tradition of mixing spirituality with science. Concerning the nature of the Art, Albertus writes:

> We must do the work ourselves, for no one can do it for us. We will begin to realize that everything is no longer as individualistic as it seemed before. We is the term in which we will think. *We*, God and I, humanity and I become entwined. The "I" loses its meaning; becomes submerged in the Cosmic All. "I" becomes many, as part of many that has its ultimate in one. Individuality, though still existing, becomes "All-individuality." Hence we begin to realize that the "I" is only a segment of the Divine, an entity in itself but not the true self, that which is All, the Divine. The wise men, sages, Adepts or whatever names we may give them, those who have become illuminated, meet on the same plane. They have climbed to the mountain top. Theirs is the mastership over the world below. They can see what happens below and that which will happen because of the far-reaching sight. Those in the valley, twisting and turning and searching behind obstacles are too close to the pattern of events to see

it. Sages read Nature as an open book printed in clear type whose sentences they fully understand.[50]

There are also 21st century practitioners of alchemy who subscribe to a more spiritual or esoteric view of the work. The Ancient and Mystical Order Rosae Crucis (AMORC) is a modern Rosicrucian Society that traces its roots to the original order going back to very early Egypt. Historians debate the existence of an actual secret society; however, it has been referenced in alchemical literature going at least as far back as the publication of the *Fama Fraternitatis* (1614) and *The Confessio* (1615) that introduced the world to the Fraternity of the Rosy Cross. Those two books were followed by *The Chemical Wedding of Christian Rosencreutz* in 1616, written by Johann Valentin Andreae. After these publications came to light, the writings became the subject of popular speculation yet some scholars such as Frances Yates,[51] believe that a mythology developed around them that had no basis in a real fraternity. Christopher McIntosh, on the other hand, discusses how alchemy became an important part of Rosicrucian ritual. He explains that the famous alchemist and proponent of the brotherhood, Michael Maier (1568-1622) of Germany:

> Believed that the Rosicrucian brotherhood had the secret of producing material gold. This secret, he maintained . . . belonged to previous civilizations and was handed down by word of mouth [*prisca sapientia*]. The Eleusinians, for example 'were very familiar with the art of making gold which they preserved and practiced so secretly that no one learned the name of the process.'[52]

Thus, in the history of the Society, it is apparent that at least some Rosicrucians believed in the physical aspect of alchemy, the creation of gold. Yet, in modern organizations, that aspect is publically downplayed whereas the spiritual side of alchemy is emphasized.

There is also a strong link between the modern Rosicrucians and the Freemasons, although the facts surrounding the connection are murky. Paul Case suggests that although Rosicrucian societies exist today, they are not the original Rosicrucian Order organized in a way as are the Freemasons. Rather, the modern Order is focused on a philosophy of obtaining self-knowledge.[53] This assertion is supported by the literature of the AMORC. Interestingly, the progress through the various Rosicrucian grades resembles Jung's individuation journey in that one strives to reach deeper contact with one's inner power as well as with the energy of the outer world.

A question arises regarding the phenomenon of joining groups such as these. What is it that compels people to search out secret societies and to participate in the complex and often, arcane rituals? The answer may be a significant reason for the spiritual-physical split that occurred in the alchemical community. The elements of the psyche that are drawn to ritual may also be more inclined toward spiritual alchemy.

The AMORC has its headquarters in San Jose, California. Its literature states:

> The Rosicrucian Order, AMORC, is a philosophical, initiatic, and traditional organization perpetuating knowledge that initiates have transmitted through the centuries. Its overall aim is to make people familiar with cosmic laws and to teach them how to live in harmony with these laws so as to achieve happiness and to acquire the mastery of life, on both the material and spiritual planes. Being neither a sect nor a religion, nor a sociopolitical movement, the Order includes in its membership men and women of all religious faiths and all stations in life.[54]

The focus of the AMORC is on developing the spiritual nature of initiates while they live a normal life. Members may choose to meet in groups but the study of the Rosicrucian teachings is accomplished at home independently. The teachings are divided into12 degrees that are studied in a progressive order. The First through the Ninth Degrees include the study of matter, consciousness, the meaning of life, cosmology, philosophy, healing, psychic abilities, their concept of God, and mysticism. Their literature does not make the last three degrees public.

The AMORC initiations can be completed alone or in groups and are said to have the following:

> Three primary objectives—first, to prepare Rosicrucian students for a new degree that they are about to study; second, to reveal to them a particular aspect of the Rosicrucian Tradition; and third, to make them listen for a few moments to the voice of the soul.[55]

The AMORC believes in reincarnation as well as developing one's psychic abilities; however, they state that the alchemy they practice in modern times is spiritual alchemy. The AMORC also suggests that "the purpose of humanity is to evolve to a unity of consciousness."[56] The Order operates in 108 year cycles of public activity and inactivity. They write that these cycles have protected them from persecution from religious or secular authorities by having these periods where they went underground. The current cycle, an active one, began in 1915 and should conclude in 2023.

Physical Alchemy versus Spiritual Alchemy

One of the critical areas of examination in my study of the women alchemists concerns the gradual rift between those alchemists who believed in a spiritual aspect to the Art as opposed to those who focused only on the physical operations. The split was not that obvious during the Middle Ages and Jung writes:

Owing to medieval ignorance both of chemistry and of psychology, and the lack of any epistemological criticism, the two concepts could easily mix, so that things that for us have no recognizable connection with one another could enter in to mutual relationship.[57]

The practice of alchemy experienced a theoretical split, very much like the Cartesian theories that emerged during the Scientific Revolution that depicted a physical distinction between spirit and matter. Physical alchemy went on to develop into the chemistry we utilize in nearly all aspects of modern life. Spiritual alchemy took a less overt path and became associated with occult studies. It will help the reader understand the implications of this change in philosophy by giving the matter/spirit split some context, beginning this story in the halls of the medieval university.

The medieval university was an amazing accomplishment in Europe. The university was divided into four faculties: arts, law, medicine and theology. A student had to complete a Master of Arts before studying the latter three subjects. The only admission requirements to the university were for students to be 14-15 years old, to pay tuition, and to take a loyalty oath. Students studied primarily with one master, forming "master-student clusters"[58] who took classes and exams together. The modern college that emphasizes the importance of teaching is not really so different from its medieval ancestor.

During the medieval period, the influx of alchemical works from the Middle East was critical in moving Western Europe out of the so-called Dark Ages. The translation of the writings of Greek and Arabic commentators on Aristotle's work, as well as other alchemical texts, was remarkable and responsible for alchemy's dissemination from the Middle East. However, alchemy was not taught in the medieval university since the curriculum focused on the *debate of theory* as opposed to *experimental, empirical practice*. Yet, alchemy was an important part of medieval medical practice and must have been passed on in some form from teacher to student.

Edward Grant provides an insightful description of the medieval curriculum of the 12th and 13th centuries that focused on the "new learning."[59] It was comprised of the Greco-Arabic writings such as those of Aristotle. Aristotle's theories made up the foundation for studying natural philosophy which focused on any physical thing that underwent change. Medieval philosophers both supported and argued against Aristotle and in 1277, the Roman Catholic Church issued a Condemnation that focused on his ideas that strayed from their accepted theology.

However, Grant notes that the medieval universities and theologians, for the most part, subscribed to Aristotle's notion of our existence. These beliefs included the argument that the earth was the center of the universe. The sun and planets orbited the earth by traversing shells in the celestial sphere that extended above the terrestrial sphere (see Ptolemy's *Almagest*, Figure 2.3). Additionally, in the Aristotelian system matter was composed of the four elements: earth, air, fire, and water and exhibited the properties of being either heavy (remained on the earth) or light (resided in the celestial realm).

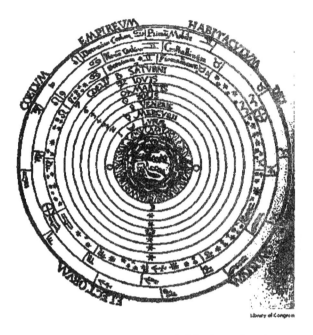

Figure 2.4: Diagram of the Planetary Orbits from the *Almagest*

Medieval natural philosophers (scientists) who were making direct observations of phenomena were in fact questioning classical writers such as Aristotle and Galen. Yet, the classical thinkers were the focus of the medieval university curriculum. The university was steeped in traditional thinking that was now being questioned by these upstart practitioners.

The Middle Ages were also a time when mystical explanations of nature flourished. Mathematics is a good example. Allen Debus explains:

> On the one hand, the new interest in mathematics furthered the development of a mathematical approach to nature and the internal development of geometry and algebra; on the other hand, the same interest resulted in occultist investigations of all kinds related to number mysticism.[60]

He adds:

> It is important *not* [emphasis mine] to try to separate the 'mystical' and the 'scientific' when they are both present in the work of a single author. To do so

would be to distort the intellectual climate of the period.[61]

The paradigm of a unified cosmos was still very strong. Debus makes the following comment which supported belief in astrology.

> Implicit in Neo-Platonism and the Christian traditions was the belief in a unity of nature, a unity that encompassed God and the angels at the one extreme and man and the terrestrial world at the other. Along with this was a continued belief in the truth of the macrocosm-microcosm relationship, the belief that man was created in the image of the great world, and that real correspondences do exist between man and the macrocosm.[62]

Natural magic is another term that came to be associated with the study of the natural world. It was not the magic that became associated with Satan and the dark arts but according to Debus, "In reality sixteenth century natural magic was a new attempt to unify nature and religion."[63] Aristotle's work was considered heretical by some Christian believers as he was thought to be an atheist. However, this put the natural philosophers, who were often proponents of Paracelsus' work, at odds with the university curriculum that taught Aristotle. For the Hermetic natural philosophers, "Science and the observation of nature were a form of divine service, a true link with divinity. In a sense natural research was a quest for God."[64] The struggle between science and religion begins to emerge in this strongly.

We see the beginning of a matter-spirit dichotomy arise in the structure of the medieval university. The arts masters stressed the value of reason as the means for generating knowledge whereas the theologians relied on divine revelation. "The arts masters ruled over the domain of reason and, therefore, of philosophy. But theologians held sway over revelation, and it is not difficult to understand why they held the upper hand in a society dominated by religion."[65] Thus, a concretization of a fundamental belief in the duality of body and spirit continued to develop.

The interdependence of science and theology is critical when considering the existence of the subsequent split between the two positions. The scientist of the Middle Ages operated from a very different paradigm than that typical of the 21st century. The scientific experiments of people with different perspectives might look similar but their conclusions about the nature of the universe will be quite different. One's intent and motivation will most certainly become the lens through which a scientist interprets observations. Marie Louise von Franz has made a significant argument that the unconscious becomes involved with the formation of new ideas.[66] As soon as the unconscious gets involved we know all sorts of archetypes, projections, complexes, and one's God-image enter the picture. They are often at odds with each other and are fodder for controversy.

Aristotle's theories stimulated debates such as whether or not the earth was eternal. Philosophers also asked whether other worlds existed beyond earth (assuming it was the center of the universe) and debated the nature of God. Concerning the three-dimensionality of God, Grant writes, "Henry More, Isaac Newton, Joseph Raphson, Samuel Clarke, and Benedict Spinoza (1632-1677) were among those who concluded that, in order to fill an infinite, three-dimensional void space, God himself had to be a three-dimensional extended being."[67] There is a sense of the *prima materia* in this thinking in that God is all pervasive, existing everywhere. This may seem simplistic to the modern reader but remember that these thinkers observed the stars as well as the empty space in-between. If God was everywhere according to Biblical teaching, that implied God was part of all matter and that idea troubled many philosophers and seemed heretical to theologians. How could someone suggest that God could be found in lower life forms?

Patai traces the split between the spiritual and material alchemists to the time of the Renaissance. He describes how the spiritual alchemists criticized practitioners whose sole goal was transmutation of base metals into gold. However, the spiritual alchemists aroused the suspicion of the Church, whose domain seemed to be usurped by these non-theological practitioners whose Art had a religious aspect and fervor. Patai writes, "One of the few to take a contrary view on this issue was Luther, who praised alchemy for its verification of Christian doctrine." [68] Jung adds that during Jakob Böehme's era, 1575-1624, a greater emphasis on the spiritual aspect of alchemy existed; however, by the 18th century, laboratory alchemy metamorphosed into modern chemistry.[69] Revelation versus reason, spirit versus matter—the stage was set for subsequent arguments and likely laid the foundation for Descartes' work although there is some disagreement regarding his true intent. This will be revisited in a subsequent chapter.

Alchemy as a Metaphor for the Development of Psyche

Alchemy as a legitimate scientific endeavor became associated with superstition and has been disregarded by many historians. With a few exceptions, if alchemical studies are mentioned in the biographies of men such as Newton, Boyle, Voltaire, and Goethe, they are done so lightly and often with a bit of embarrassment on behalf of the historical figure. Furthermore, the alchemy discussed is usually of the more practical or physical nature as opposed to the spiritual aspect of the Art. Jung, however, observed a different way to look at the practice of alchemy that sheds light on the reason such respected scientists found alchemy to be so vital. Jung perceived connections between psychological development and alchemy that others had not noticed. In alchemy, Jung found a parallel to his theory of the development of the psyche. This was no easy task and Jung writes that making sense of alchemical writing was laborious. He comments on the beginning of his own work writing:

> Occasionally I would look at the pictures, and each time I would think, 'Good Lord, what nonsense! This stuff is impossible to understand.' But it persistently

intrigued me . . . occasionally I even found a few sentences which I thought I could understand. Finally, I realized that the alchemists were talking in symbols.[70]

Jung also noticed the link between alchemy and Gnostic beliefs: "Grounded in the natural philosophy of the Middle Ages, alchemy formed the bridge on the one hand into the past, to Gnosticism, and on the other into the future, to the modern psychology of the unconscious."[71] Keep in mind, Newton as well as his peers were thinkers of their time. Their science was enveloped by their world view and even more critically, by their theology. Whichever theories they developed had to be in line with their religious beliefs. Anything that hinted at atheism would be discarded and dismissed. Jung believed that Gnostic writing contained symbolism that derived from the unconscious. He would likewise go on to make the connection between alchemy and the unconscious.

Jung made a remarkable observation regarding alchemy and individuation or psychic development; he noticed that the process that an individual undergoes in the course of the individuation journey was mirrored in the alchemical operations. Jung had been interested in alchemy for some time when he noticed that his patients reported dream imagery that brought to mind alchemical symbolism although they knew nothing of alchemy. Jung concluded that dream imagery provides clues regarding the progress of a person's individuation journey, a process he believed that the psyche is compelled to undergo for psychological growth. In *Psychology and Alchemy*, Jung explains that both individuation and alchemy consist of steps in which the matter being worked upon undergoes calcination, distillation, separation, putrefaction, and the final union or *coniunctio*. That is, when the Self individuates, the goal is for the opposites, the conscious and the unconscious, to come into relationship with each other, to form a union. The union of the opposites is also the goal of the alchemical process being characterized by the *coniunctio* phase, in which the elements, often symbolized by Sol and Luna or male and female, are recombined, creating the Philosopher's Stone. This will be discussed in more detail in my summary of Edward Edinger's work; however, it should be noted that Jung is explicit in his description of the close association between alchemy and individuation. For example, the *prima materia* correlates to the undifferentiated unconscious before individuation begins; both are in a state of chaos. The work takes place in the alchemical vessel as well as in the psyche. Even the alchemical vessel, which has a round or egg shape, is said to mirror the structure of the universe, as well as the uterus.

According to Jung, the alchemists projected unconscious material into their understanding of the work they were doing.

Everything unknown and empty is filled with psychological projection; it is as if the investigator's own psychic background were mirrored in the darkness. What he sees in matter, or thinks he can see, is chiefly the data of his own unconscious which he is projecting into it.[72]

Jung was not suggesting that alchemy was imaginary as some science historians have concluded when reading his work. He was explaining the way alchemists understood their practice. This was early chemistry and the alchemists were trying to make sense of the unknown. We moderns have withdrawn these projections, which is why we no longer identify personally with chemical reactions as though they were mythological events. Jung maintained that the withdrawal of projections occurs when the projections become conscious. This process of making projections conscious and withdrawing them from their object is a central focus of psychotherapy.

Alchemy, at least until the spirit-matter split became so concrete, was a physical operation as well as a symbolic effort. Jung asked us to consider why an alchemist would work in the laboratory if his work was *only* symbolic. On the other hand, why did the alchemists use such obscure and symbolic language to describe the process? Their directions did very little to help the reader gain a clear understanding of the course of the work. Jung writes:

> I am therefore inclined to assume that the real root of alchemy is to be sought less in philosophical doctrines than in the projections of individual investigators. I mean by this that while working on his chemical experiments the operator had certain psychic experiences which appeared to him as the particular behaviour of the chemical process. Since it was a question of projection, he was naturally unconscious of the fact that the experience had nothing to do with matter itself (that is, with matter as we know it today). He experienced his projection as a property of matter; but what he was in reality experiencing was his own unconscious.[73]

The alchemists believed that the projection was a real phenomenon. How better to describe the combining of substances than as a sacred marriage! Jung's points emerge from the theme that will keep appearing in this book, that alchemy was both symbolic and practical and thus, was manifested in different types of alchemical practice. The work required one to measure solutions, dry ingredients, soak matter, and distill the *menstruum*[74] as well as to settle one's mind and to meditate upon the transformations taking place. Furthermore, the alchemists believed that the operations taking place in the vessel were also taking place in their own bodies. Although some alchemists were aware of and recorded visions, they would not have understood their genesis in the unconscious. The concept of the unconscious did not become part of formal psychological theory until Freud named it as such. Yet, the alchemists did realize that their imagination played a critical role in the work. Jung writes in *Alchemical Studies*:

> For what in the end, do we know about the causes and motives that prompted man, for more than a thousand years, to believe in that "absurdity" the transmutation of metals and the simultaneous psychic transformation of the artifex? We have never seriously considered the fact that for the medieval investigator the redemption of the world by God's son and the transubstantiation of the Eucha-

ristic elements were not the last word, or rather, not the last answer to the manifold enigmas of man and his soul. If the *opus alchymicum* claimed equality
with the *opus divinum* of the Mass, the reason for this was not grotesque pre-
sumption but the fact that a vast, unknown Nature, disregarded by the eternal
verities of the Church, was imperiously demanding recognition and acceptance.
Paracelsus knew, in advance of modern times, that this Nature was not only
chemical and physical but also psychic . . . . And even if, like all the rest of
them, he never produced any gold, he was yet on the track of a process of psychic transformation that is incomparably more important for the happiness of
the individual than the possession of the red tincture.[75]

As I mentioned, the goal of individuation is the unification of the ego with
the unconscious in which the ego comes into relationship with the unconscious,
a process entailing what Jung labeled the *transcendent function.*

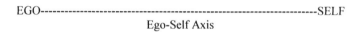

EGO---SELF
Ego-Self Axis

Jung argued that the individuation process results in the realization of the Self,
the totality of the psyche. Jeffrey Raff describes different levels of Self-
formation. The first level is characterized by the ego becoming aware of and
exploring the unconscious. The ego has to realize that there is something beyond
itself, that it might be part of a greater whole. The second level is called Self-
formation; however it is the third level I wish to emphasize. The third level
moves awareness into a place where "the individual self that has been formed
comes into union with a level of reality that transcends it, with the divine world
that Dorn (an alchemical/philosophical writer) called the '*unus mundus.* "[76] The
third level is a state of awareness of a realm outside daily existence, often referred to by depth psychologists as the *psychoid.* Raff discusses this psychoidal
realm, which differs from the imaginal in that the images that appear to the individual are thought to be from outside the unconscious rather than from within.
This realm may be closer to what the alchemists sensed as coming into contact
with God. It is a place of union with all reality.

Jung correlated the Self with the God-image. Thus, the individuation process brings us into relationship with our image of the Divine. In other words, the
seeking of Self is a search for God. So-called *true* alchemists were searching for
God as well as the Philosopher's Stone. Raff adds:

The symbols of alchemy depict actual experiential states and processes. Alchemy is not just a metaphor for psychological life, but the insightful portrayal
of imaginal experiences that create profound transformation in the psyche . . . .
Alchemy is a not just a guide for inner work, but a map of the way.[77]

Studying alchemy gives us a wonderful opportunity to glimpse the nature of
the mind maps made by, not only the early alchemists, but also scientists and

philosophers such as Spinoza and Newton. It is as if the transcendent function creates the psychic space for sacred events to be experienced. If this is so, the alchemist was in a strong position to experience the transcendent function. As von Franz explained, working in isolation, trying to understand the nature of the operations, and having a sense that one was doing sacred work opened the way for the alchemists to participate in a spiritual practice, not just a purely practical one. Within this context, the complex writings of the early alchemists make more sense as they are expressions of a psychic process. In the same way that the unconscious expresses itself in mythology, it could also make itself known in alchemical symbolism.

Von Franz explained that alchemical symbolism is important to our understanding the unconscious because it was not diluted by religion or other traditions. She notes that when ideas are communicated between generations, views that seem unimportant are often left out:

> This is what happens to original experiences which are handed on, for a selection is made and what fits or coincides with what is already known is handed on, while other details tend to get dropped, because they seem strange and one does not know how to deal with them.[78]

The alchemists simply observed unfamiliar phenomena and described it in terms that were more congruent with their experience. Von Franz describes the alchemist as being, "respected, but also hated and feared and therefore such things had to be done secretly and in loneliness, which always brings up the unconscious."[79]

Von Franz also asks why modern thinkers would be interested in studying alchemical symbolism. She compliments Jung when she explains that:

> One of the great merits of Jung is that he gave us a key to these texts which are looked upon officially by historians of chemistry as absolute nonsense, for to them they mean nothing at all. But to us it is clear what Olympiodoros [5th century alchemical writer] is driving at, namely an inner experience, an introverted religious experience which those people had in their meditations and experiments with material phenomena. That was the basis of alchemy.[80]

In other words, we see the nexus between spiritual and material alchemy. Yet, science historians and depth psychologists continue to struggle with the idea that a nexus even exists. Historians look at alchemy as a precursor to modern chemistry. Depth psychologists tend to spend most of their time in alchemy's spiritual realm.

Besides giving us a sense of the history of scientific evolution, von Franz argues that alchemy illustrates what still happens when we try to explain the unknown.

Great discoveries in natural sciences are generally due to the appearance of a new archetypal model by which reality can be described; that usually precedes big developments, for there is now a model which enables a much fuller explanation than was hitherto possible.[81]

That new model is held as truth until a new projection is made resulting in yet a newer model. Von Franz continues by examining how the unconscious exerts influence in making these projections about our reality. She poses a rather brilliant question:

But what role *does* the unconscious play [in scientific work]? The unconscious seems to deliver models which can be arrived at directly from within without looking at outer facts, and which afterwards seem to fit outer reality. Is that a miracle or not? There are two possible explanations: either the unconscious knows about other realities, or what we call the unconscious is a part of the same thing as outer reality, for we do not know how the unconscious is linked with matter. If a wonderful idea as to how to explain gravitation comes up from within me, can I say that it is the nonmaterial unconscious giving me a wonderful idea about material reality, or should I say that the unconscious gives me such a marvelous idea of outer reality [emphasis mine] *because it is itself linked with matter, it is a phenomenon of matter and matter knows matter?*[82]

Allowing for the unconscious to participate in ego-driven science may explain why some historians of science do not care for Jungian thinking in alchemical studies. The Jungians do not negate the physical or empirical side of research; they just remind us that there are unconscious aspects of the psyche that are always in play regardless of the emphasis some scientists place on objectivity and compartmentalization.

Edward Edinger on Jung and Alchemy

Edward Edinger discussed Jung's alchemical metaphor in great detail in his seminal book, *Anatomy of the Psyche.*[83] Like von Franz, he wrote that Jung's study of alchemy illustrated psychic processes. Noting the parallels between alchemy, individuation, and psychotherapy, Edinger writes, "What makes alchemy so valuable for psychotherapy is that its images concretize the experiences of transformation that one undergoes in psychotherapy."[84] He reminds us of the spiritual aspect of alchemy that is mirrored in the process of searching for the Self. Just as the alchemist must carry out the necessary operations to achieve the Philosopher's Stone, so the ego is needed to "cooperate deliberately in the task of creating consciousness."[85] Individuation does not happen in a passive environment, but like the turmoil in the alchemical vessel, it involves work and pain on the part of the pilgrim.

What makes Edinger so useful to read is the way he links physical alchemy with the spiritual/psychological metaphor. He takes each operation beginning with the laboratory process and then adds the psychic interpretation. Thus, it becomes possible to glimpse the way the alchemists carried out their practical work while describing it in a metaphysical way. I will summarize Edinger's discussion of each operation very briefly. I encourage the reader to spend time with *Anatomy and Psyche*, a pivotal work and one that can be read repeatedly.

Edinger describes the *prima materia* writing that "the world derives from a single, original stuff, the so-called first matter."[86] This sounds very similar to the first substance described by Michael Sendivogius and we will see this again when we look at the connection between alchemy and the Kabbalah. Edinger echoes Jung as he compares the undifferentiated unconscious to the basic, *prima materia*. Transmutation of matter requires the alchemist to simplify matter to the *prima materia*; psychotherapy serves a similar function. "The fixed, settled aspects of the personality that are rigid and static are reduced or led back to their original, undifferentiated condition as part of the process of psychic transformation."[87] Edinger describes this state as chaotic which is also the description Sendivogius gives to first matter. The feeling of being in this chaotic, primal place is not particularly pleasant but is a necessary step in our psychic work.

Edinger amplified Jung's work as he explained the individuation process in the context of each alchemical step. He begins with the operation *calcinatio* in which the purification of the metal or other substance is compared to the purification or purging of unhealthy psychic contents. The impetus for *calcinatio* typically comes from the energy that surrounds frustrated desire and may either follow or precede *solutio*, the next step. Calcining a substance reduces it to ash, which is often how a person feels who is in this phase of individuation.

Solutio's psychic equivalent to dissolving matter in a solvent is the process of the ego descending into unconscious material. In dream imagery we often experience *solutio* as swimming in the ocean.

> Whatever is larger and more comprehensive than the ego threatens to dissolve it. Internally, the unconscious as the latent Self or totality of the psyche can dissolve the ego. Externally, an individual with a larger consciousness than one's own can bring about *solutio*.[88]

The alchemists dissolved matter in water, urine, or any liquid that was believed to successfully return the sulphur, mercury, and salt to solution. The *solutio* in psychotherapy occurs during the transference between patient and therapist. Hopefully, this is a containing experience where the therapist helps hold the material that is being worked upon by the patient. However, there are times in the therapeutic work when the analysand feels awash like a leaky boat. Fortunately, the therapist can assist with mending the container resulting in it being stronger than it was before the process began.

The next step, *coagulatio*, occurs when matter precipitates out of solution. For example, if copper is in solution, one can extract it by adding lime or sodium hydroxide.[89] Edinger characterizes this as a solidifying experience for the psyche, which occurs in a state of disorganization as opposed to one of calm. By becoming attached to the ego, the material can be discerned and differentiated, which allows consideration of its place within the context of the entire psyche. One needs to be able to hold the material psychically, turn it over, and observe it from all angles in order to make sense of how it came to be in this place. When issues are residing solely in the unconscious, they are ethereal and immune to reflection.

Sublimatio occurs as the spirit of the alchemical matter rises in the flask. Psychologically, Edinger explains the dissociative aspect of *sublimatio*. "The capacity to get above and see oneself objectively is the ability to dissociate. The use of this word immediately indicates the danger of *sublimatio*. Each of the alchemical operations has its own pathological symptomatology when carried to extremes."[90] Edinger distinguishes between a lesser and greater *sublimatio* in that the latter is climactic, whereas "the lesser *sublimatio* must always be followed by a descent."[91] This is akin to heating the contents of the alchemical vessel for a long period of time, allowing the contents to vaporize and condense within the sealed flask in a circulatory pattern.

The descent to the unconscious contributes to the psychological circulatory pattern found in this phase. This provides a space for a balancing aspect to the operation which is typically repeated many times.

> One must make the circuit of one's complexes again and again in the course of their transformation. The 'powers of the above and the below' are combined to the extent that the unified personality is created that connects the personal psyche (below) with the archetypal psyche (above).[92]

Thus, we may experience a descent to the Underworld, work though the complex, and feel as if we have it under control, only to have an outer event precipitate further work.

There is a phase in alchemy when the matter goes through a decaying and death, *mortificatio*. "*Mortificatio* is the most negative operation in alchemy. It has to do with darkness, defeat, torture, mutilation, death, and rotting. However, these dark images often lead over to highly positive ones—growth, resurrection, rebirth."[93] Edinger explains that this phase is all about gaining awareness of one's shadow, the darker, unconscious aspect of the psyche. It feels like a death in which old ideas that are not working are put to rest, opening the way for a rebirth and an emergence of new attitudes. A universal constant in fairy tales is the experience of a real or metaphorical death by the hero or heroine. No matter how the death comes about, it must happen before the next step can occur.

Mortificatio is followed by *separatio*. Once the matter has decayed, it is put back into solution and distilled to separate the pure matter from the contaminated contents. Psychologically, *separatio* is a growing sense of the opposite qualities in the psyche. The idea of opposites has to do with the antagonistic attitudes found within our psyches. For example, people generally hold a mixture of atti-

tudes that are in opposition to each other. The holy man covets material posses-
sions; the chaste young woman thinks thoughts she feels are impure. Their behavior is normal although many religious frameworks call for repressing the
attitudes deemed negative or *bad*. Depth psychologists are well aware of the
danger that accompanies a refusal to allow unconscious attitudes to be acknowledged although not necessarily acted upon. Repression can open a space for
people with weaker egos to be dominated by others who may or may not have
their best interests in mind. They may also project their shadow images onto
others resulting in irrational feelings, racism, hate crimes and in one of the more
infamous instances, the *Shoah* and Hitler's death camps.

> To the extent that the opposites remain unconscious and unseparated, one lives
> in a state of *participation mystique*, which means that one identifies with one
> side of a pair of opposites and projects its contrary as an enemy . . . one be-
> comes conscious as one is able to contain and endure the opposites within.[94]

The word *endure* is an apt description of what it can sometimes feel like to
hold opposing ideas in balance. Eastern philosophy addresses the need to bring
balance between adversarial attitudes in the creation of consciousness. For this
balance to occur, one must be able to discern the two opposite attitudes and that
is why *separatio* is so important. Edinger reminds us of the *separatio* that occurs
in a variety of Creation myths where the world is formed by separation of earth
and sky. He also discusses the fact that both religion and philosophy differenti-
ate spirit and matter. The split between the two seems to have an archetypal root
in that the soul leaves the body upon death. "*Separatio* is closely connected to
the symbolism of *mortificatio*, which means *separatio* may be experienced as
death."[95] The motif of *separatio* is seen in the spirit-matter split that is one of the
topics of this book. This will be discussed in more detail in a subsequent chapter
as it pertains to the spirit-matter split of the seventeenth century. In alchemy,
however, *separatio* is not the end of the work. Alchemically, the feminine and
masculine principles must be purified separately before they can be joined in
coniunctio.

> They must be thoroughly cleansed from contamination with each other, which
> means diligent and prolonged scrutiny of one's complexes. When the *separatio*
> is completed, the purified opposites can be reconciled in the *coniunctio*, which
> is the goal of the *opus*.[96]

The alchemists observed that the combination of substances formed a new, third
substance, also called the Philosophical Child or Philosopher's Stone. It sounds
rather simple, but psychologically, the *coniunctio* phase has two steps: the lesser
and greater *coniunctio*. Edinger explains:

> The lesser *coniunctio* is a union or fusion of substances that are not yet thor-
> oughly separated or discriminated. It is always followed by death or *mortifica-*

tio. The greater *coniunctio*, on the other hand, is the goal of the *opus*, the supreme accomplishment.[97]

Regarding the lesser *coniunctio*, Edinger states that during analysis, as unconscious material comes to light, the ego may identify with it. This might take the
form of identification with the shadow. He adds that on an outer level, one is in
the lesser *coniunctio* when one becomes over-identified with groups or individuals. In both cases, "such contaminated *coniunctio* must be followed by *mortificatio* and further *separatio*."[98]

The greater *coniunctio* is analogous to the formation of the Philosopher's
Stone or psychologically, knowing the Self. The polar opposites that have
achieved a pure state are finally unified and one is said to be able to hold the
tension of the opposites, in other words, to achieve consciousness. Jung considered the Philosopher's Stone a Christ symbol with a very interesting difference.
In most Christian traditions, Christ does not represent the union of opposites, as
he is all righteousness, lacking shadow. If the goal of individuation is the *coniunctio*, then we need to achieve a union of good and its opposite, evil. Thus, the
complete Christ must have a shadow side that he understands, relates to, and
does not allow to overwhelm his ego. Jung wrote about this in *Answer to Job*
(1958/1969).[99] This poses a troubling question for Christianity and will be discussed in another section. However, Edinger summarizes the individuation-
alchemy link beautifully.

> The psychological rule is: the unconscious takes the same attitude toward the
> ego as the ego takes toward it. If one pays friendly attention to the unconscious
> it becomes helpful to the ego. Gradually the realization dawns that a mutual
> *opus* is being performed. The ego needs the guidance and direction of the un-
> conscious to have a meaningful life; and the latent Philosopher's Stone, impris-
> oned in the *prima materia*, needs the devoted efforts of the conscious ego to
> come into actuality. Together they work on the Great Magistry to create more
> and more consciousness in the universe.[100]

Consequently, although the individual does psychic work for a lifetime,
there is a profound peace that comes with self-understanding and self-
acceptance. In order to overcome one's complexes, the *coniunctio* phase reminds us that we are not perfect beings, that we have choices, and that we can
even come to have compassion for those more unsavory parts of ourselves that
respond better to a loving hand rather than the rod.

Kabbalah and Alchemy

Alchemists were often well acquainted with the study of the Kabbalah. Kabbalah refers to the framework for Jewish mysticism in which the world is made

manifest from God who is eternal and without limits.[101] Z'ev Shimon Halevi explains that Kabbalah is as old as time.[102]

> Kabbalah, it is said, goes back to the angels who were instructed by God. Mankind is taught by the chief archangel Metatron, who Apocalyptic Legend says is the transfigured Enoch, the man who walked with God and did not taste death (Genesis 5:24).[103]

Helevi also relates the account that Kabbalah was handed down through the Jewish patriarchs beginning with Abraham through Isaac, Jacob, Levi, Moses and on to Joshua, David and Solomon. Solomon broke the line of teaching by worshipping other Gods and the Temple was destroyed. The Jews were exiled to Babylon where Kabbalah went underground. It resurfaced at a later time but was studied quietly, taking a back seat to the study of Talmud and the Torah.[104] It is believed that Christ studied this work as well. The Kabbalah came to Western consciousness surreptitiously, trickling into Europe slowly. *The Zohar* or *Book of Splendour* attributed to Moses de Leon, surfaced around the 13th century. Isaac is also an often cited expert as he wrote much on the Kabbalah during the 16th century.

The Kabbalah was taken very seriously by many scholars as well as being linked to those who would misappropriate the teachings into their practice of magic. Arthur Waite, a well-known 19th century scholar of occult studies was especially critical of people who associated Kabbalah and Ceremonial Magic.[105] Waite discusses certain elements of Kabbalah reluctantly but states that he feels it is necessary due to its popularity with occultists. For example, discussion of the divinity of numbers can be found in both serious and faddish writings. Each number represents concepts regarding the universe such as:

- Cornelius Agrippa [alchemist] associated 10s = celestial
- 100s = earthly
- 1000s = things to come

As well, if one wanted to compute the value for the Name of God known as the Tetragrammaton it would look like the following. Each Hebrew letter that makes up the name, Yahweh, has a numerical value. The computation involves adding the values of the letters in increments. Added together one arrived at the sacred number 72.

Jod = 10
Jod He = 15
Jod He Vau = 21
Joh He Vau He = 26
10+15+21+26=72

Kabbalah is a complex, profound system that explains the nature of the universe. Thus, it is logical that alchemists would be interested in what they found were many parallels between the two schemata. For example, in a manner very similar to the alchemical *prima materia*, the first matter of Kabbalah contains all that there is in the universe which is of God and is manifested as our world. Halevi explains:

> Tradition states that God willed to see God and so God's Will, symbolized by light, shone nowhere and everywhere. Thus Tradition states that God willed to see God and so God's Will, symbolized by light, shone nowhere and everywhere. Thus the EN SOF AUR, the Endless Light of Will, was omniscient throughout Absolute All. From God knowing All, God willed the first separation so that God might behold God. This, we are told, was accomplished by a contraction in Absolute All, so as to make a place wherein the mirror of Existence might manifest . . . . The first manifestation at the circumference of the void was named the Prime Crown. It has many other titles, like the Concealer of the Concealed, the White Head and the Crown of Crowns. Most Kabbalists knew it by the God Name of EHYEH or I AM, where the Absolute allowed Existence to be . . . . When God willed the World to come into being, the seed took root and grew downward into the trunk, branch and fruit of a Divine Tree that would act as an intermediary between the World and God.[106]

Thus, the motif of the world being born out of chaos is also seen in Kabbalah. The interconnectedness of matter is described. The operation of *separatio* is mentioned which allows the aspects of God to become self-aware. It is no wonder that the alchemists studied this mysticism as it paralleled their work so well. We will return to the parallels between alchemy and the Kabbalah in the discussion of Dorothy Moore and Katherine Jones, Lady Ranelagh and their studies in Millenarianism.

This chapter has covered a lot of information while barely skimming the surface of a subject that has been explored, analyzed, derided and puzzled over by scholars for thousands of years. It is my hope that the reader can see from the topics I have addressed in this chapter that although Descartes may have thought he separated spirit and matter, they remain inextricably connected in at least, the unconscious. Furthermore, although an experimenter might deny the role the unconscious plays in research, it keeps manifesting in one of the goals of science, to discern the nature of our existence. We will see that the women alchemists may have thought they were able to employ the new objective science; however, one cannot prevent the unconscious from affecting one's thinking.

Notes

1. See Martinus Rulandus, *A Lexicon of Alchemy Or Alchemical Dictionary* (Frankfurt: Zachariah Palthenus, 161. Reprinted by Montana, U.S.A.: Kessinger Publishing),

409. The Lexicon covers a copious amount of terms that students of alchemy will encounter in reading older tracts.

2. Ralph Patai, *The Jewish Alchemists: A History and Source Book* (Princeton, NJ: Princeton University Press, 1994), 3.

3. Henry Cornelius Agrippa, *Fourth Book of Occult Philosophy* (London: Askin, 1978/1655), A3.

4. H. Stanley Redgrove, *Alchemy: Ancient and Modern* (New Hyde Park, NY: University Books,1969); C. J. S Thompson, *Alchemy and Alchemists* (Mineola, NY: Dover, 1932/2002); Marie Louise Von Franz, *Alchemy: An Introduction to the Symbolism and the Psychology* (Toronto: Inner City Books, 1980).

5. Michael White, *Isaac Newton: The Last Sorcerer* (Reading, MA: Persues Books, 1997), 4.

6. Masumi Chikashige, *Alchemy and Other Chemical Achievements of the Ancient Orient: The Civilization of Japan and China in Early Times as See from the Chemical Point of View* (Tokyo: Rokakuho Uchida, 1936).

7. Robert P. Multhauf, *The Origins of Chemistry* (New York: Franklin Watts, 1966).

8. Multhauf, 121.

9. Patai, 4.

10. Note the similarity between this description and that of the currently identified Higgs-Boson particle!

11. Von Franz, 67.

12. Reinhard Federmann, *The Royal Art of Alchemy*, trans. R. H.Weber (Philadelphia: Chilton, 1964/1969), 30.

13. Federmann, 31-32.

14. Frater Albertus, *The Alchemist's Handbook: Manual for Practical Laboratory Alchemy* (York Beach, Maine: Samuel Weiser, 1974).

15. Waite, 5-6.

16. Albertus, 123.

17. White, 108

18. Redgrove, 41.

19. Thompson, 26.

20. Thompson (1932/2002).

21. Patai, 66.

22. Thompson, 61.

23. Fernando (1998).

24. Michael Sendivogius, *The New Chemical Light*, nd.

25. Thompson (1932/2002).

26. Waite, 79.

27. Waite, 79.

28. Sendivogius, 81.

29. Sendivogius, 96.

30. Sendivogius, 101-102.

31. Carl Jung, *Alchemical Studies* (Princeton, NJ: Princeton University Press, 1967), 120.

32. Carl G. Jung (1967); Henry Sigerist (1941).

33. Thompson, 168.

34. J. H. Oxen, trans., *Paracelsus His Archidoxis, Or, Chief Teachings; Compiled in Ten Books, Disclosing the Genuine Way of Making Quintessences, Arcanums, Magisteries, Elixirs, &c.* (London: Lodowick Lloyd, 1663), 61.

35. Jung, 120.

36. C. Lilian Temkin, trans., "Seven Defensiones, the Reply to Certain Calumniations of His enemies." In *Paracelsus: Four Treatises*, edited by Henry E. Sigerist (Baltimore, MD: Johns Hopkins University Press, 1941), 13-14.

37. Temkin, 18.

38. Temkin, 24.

39. Temkin, 31.

40. Temkin, 34.

41. Temkin, 41.

42. Jung, 114.

43. Jung, 118.

44. See Michael Maier, *A Subtle Allegory Concerning the Secrets of Alchemy Very Useful to Possess and Pleasant to Read.* Edmonds, WA: The Alchemical Press, 1984. Reprint; Lynn Thorndike. *A History of Magic and Experimental Science, volumes VII and VIII: The Seventeenth Century* (New York: Columbia University Press, 195), 8.

45. Maier, 9.

46. Manfred Junius, *The Practical Handbook of Plant Alchemy* (Rochester, VT: Healing Arts Press, 1993).

47. Frater Albertus (1974).

48. Mark Stavish, *The Path of Alchemy* (Woodbury, ME: Llewellyn, 2006).

49. Stavish, 4.

50. Albertus, 21.

51. Frances Yates, *The Rosicrucian Enlightenment* (London: Routledge, 1972).

52. Christopher McIntosh, *The Rosicrucians: The History, Mythology, and Rituals of the Esoteric Order* (York Beach, MA: Samuel Weiser, 1997), 33-34.

53. Paul Foster Case, *The True and Invisible Rosicrucian Order: An Interpretation of the Rosicrucian Allegory and an Explanation of the Ten Rosicrucian Grades* (York Beach, ME: Samuel Weiser, 1981).

54. Christian Bernard, F. R. C., ed., *Rosicrucian Order AMORC: Questions and Answers* (San Jose, CA: Supreme Grand Lodge of AMORC, Inc., 1996/2001), 13.

55. Bernard, 41.

56. Bernard, 81.

57. Jung (1963/1970), 467.

58. Edward Grant, *The Foundations of Modern Science in the Middle Ages* (Cambridge: Cambridge University Press, 1996), 39.

59. Grant, 43.

60. Debus, Allen G. *Man and Nature in the Renaissance* (Cambridge: Cambridge University Press, 1978), 11.

61. Debus, 11.

62. Debus, 13.

63. Debus, 13.

64. Debus, 14.

65. Grant, 72.

66. Von Franz (1980).

67. Grant, 126.

68. Patai, 4.

69. Carl G. Jung, *Psychology and Alchemy*, 2nd ed. (Princeton, NJ: Princeton University Press, 1953/1968).

70. Carl G. Jung, *Memories, Dreams, Reflections* (New York: Vintage Books, 1961), 204.

71. Jung, 201.

72. Jung (1953/1968), 228.

73. Jung, 245.

74. The menstruum refers to the liquid portion of the alchemical work.

75. Carl G. Jung, *Alchemical Studies* (Princeton, NJ: Princeton University Press, 1967), 159-160.

76. Jeffrey Raff, *Jung and the Alchemical Imagination* (York Beach, ME: Nicolas-Hays, 2000), 85.

77. Raff, 253.

78. Von Franz, 16.

79. Von Franz, 78.

80. Von Franz, 90.

81. Von Franz, 33.

82. Von Franz, 36-37.

83. Edward F. Edinger, *Anatomy of the Psyche: Alchemical Symbolism in Psychotherapy* (Chicago: Open Court, 1985).

84. Edinger, 2.

85. Edinger, 8.

86. Edinger, 10.

87 Edinger, 10.

88. Edinger, 56-57.

89. Edinger, 83.

90. Edinger, 126.

91. Edinger, 140.

92. Edinger, 144.

93. Edinger, 148.

94. Edinger, 187.

95. Edinger, 202.

96. Edinger, 209.

97. Edinger, 211.

98. Edinger, 215.

99. See Carl G. Jung, *Answer to Job*, translated by R.F.C. Hull (Princeton: Princeton University Press, 1958/1969).

100. Edinger, 230.

101. Epstein (1988); Halevi (1976, 1977).

102. Z'ev Ben Shimon Halevi. *The Way of Kabbalah* (York Beach, ME: Samuel Weiser, 1976).

103. Halevi, 16.

104. Halevi (1976).

105 Arthur E. Waite, *The Holy Kabbalah* (New Hyde Park, NY: University Books, 1960). (Work originally published in 1902 as The Doctrine and Literature of the Kabbalah with Theosophical Society in London as publisher)

106. Halevi, 8-9.

Chapter 3

The Feminine Presence in the Magnum Opus

I make the operations easy, and explain myself the most clearly that it may be, for them to learn how to do themselves the things of which they will have need. From the foreword of Part Six of *La Chymie charitable et facile, en faveur des Dames Charitable* [*Easy Chemistry for Women*], 1656 by Marie Meurdrac.

Walter Endrei's 1974 *Old Chemical Symbols* displays an intriguing print titled, "Distillation."[1] It was attributed to Philipp Galle (1537-1612) but was based on a painting by Jan van der Straet titled, "Distillation, Boiling Water to Purify It." A 17th century alchemist's lab is portrayed showing numerous men working at the furnace, carrying a retort, reading an old manuscript, and discussing the task at hand. Two young boys are also employed. One is holding a large pestle that is suspended from a metal pole, which allows him to grind what looks like wheat in the mortar. The other boy is using a small bellows to stoke the fire in the furnace or athanor. The alchemist is referring to a book, seemingly with an assistant or colleague looking over his shoulder. There is the usual hustle and bustle one learns to expect in a laboratory.

None of this is unusual. However, if one looks closer, there is an archway that leads to another room that shows a woman seated before what appears to be a large stone table. It is difficult to tell for certain but either there is a fire behind her or it is burning in a depression in the center of the table. Who is she? Her presence is not an afterthought although her place is in the background. That she is an important part of the work seems clear but her role is uncertain and that is one of the best ways to describe the work of the early women alchemists.

The women who are the focus of this chapter have been given more recognition than many of the other women in this book. Yet, there is relatively little known about their lives and their work.

Maria Hebraea: Inventor

A young Jewish woman rushes toward her old mentor shouting, "'One becomes two, two becomes three, and by means of the third and fourth achieves unity; thus two are but one.'"[2] Most alchemists believed that she was describing successful transmutation and unlocking the meaning of her somewhat ambiguous axiom would be akin to finding the key to creating the Philosopher's Stone. Carl Jung analyzed Maria's axiom, but its precise meaning remains an alchemical mystery. The woman is Maria Prophitissa or as some call her, Maria Hebraea or Maria the Jewess (see the cover of this book). The story of the women alchemists does not belong solely to Maria the Jewess, but her work is a helpful place to set about discussing women's role in the alchemical tradition.

Zosimos (5th century CE) wrote profusely about alchemy, providing early records of practical information such as the description of early laboratory equipment.[3] Zosimos also described the prolific Jewish tradition in alchemy. He related the story of Maria, a young Jewish woman who lived in Alexandria around the first century CE. There is scant information available about Maria; however, she is recognized as having written one treatise, *Maria Practica*. Maria is also credited with inventing or, at least making famous, many kinds of apparatus used for distillation and sublimation, critical alchemical operations. The *balneum mariae* or:

> 'Maria's bath' resembled a double-boiler and was used, as is a modern water bath, to heat a substance slowly or to maintain it at a constant temperature. In modern French a double-boiler is still referred to as a *bain-marie*.'[4]

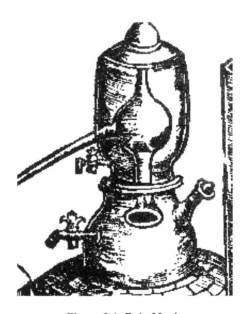

Figure 3.1: Bain Marie

Maria's inventiveness is also believed to include the *tribikos* which was a type of still and the *kerotakis*.

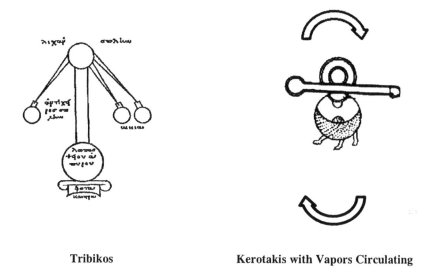

Tribikos   Kerotakis with Vapors Circulating

Figure 3.2: Tribikos and Kerotakis

The *kerotakis* was a vessel that allowed Maria to experiment with the effect of the vapors of mercury, arsenic, and sulphur on other metals in the *sublimatio* operation of alchemy. It was constructed in a manner in which the element was heated to boiling so that the vapors given off rose to the top of the sealed apparatus, condensed, and proceeded to bathe the metal being experimented upon. The process might be visualized as bathing in one's own vaporized essence. As the substances being heated reached their boiling point, the vapor that would rise would collect at the top of the retort, the vessel used in the work. Later, the alchemist would find that the vapor had left a salt in the vessel that dried in the form of crystals. The conclusion was that the spirit of the substance being heated had united with some heavenly spirit. Ronald Gray describes so nicely the way that *sublimatio* seemed to work that it might clarify further the type of work in which Maria engaged.

> The 'elements,' or minerals, were placed within a closed vessel and there heated. As boiling point was reached, some of the contents evaporated and rose upward in the form of vapour or gas. In this way, the spirit of the metals was supposed to have been released from its base surroundings, and it was this spirit which was said to be rising within the vessel. In doing so, however, it frequently formed a deposit on the inner surface in the shape of flower-like crystals.

These crystals were said to be a salt of great power, which had been formed by the union of a higher being with the escaping spirit of the metals [5]

Leonora Carrington, famed surrealist, painted an extraordinary depiction of Maria carrying out transmutation or as it was also termed, *Chrysopeia*. The work is titled, *The Chrysopeia of Mary the Jewess*. I find Leonora Carrington's vision of Maria mesmerizing and could not stop thinking about it in the context of this book. I will come back to this painting in chapter eight but it is remarkable in its unification of several concepts from alchemical practice as well as from the Kabbalah. Maria Prophitissa is one of the more well-known and accepted women alchemists but the reader will see that most of these women were not thought of so highly.

Hypatia of Alexandria: Her Experience of *Mortificatio*

Margaret Alic quotes Socrates Scholasticus, who describes a terrible murder that took place in March 415 CE.

> 'All men did both reverence and had her [Hypatia] in admiration for the singu-lar modesty of her mind. Wherefore she had great spite and envy owed unto her, and because she conferred oft, and had great familiarity with Orestes, the people charged her that she was the cause why the bishop and Orestes were not become friends. To be short, certain heady and rash cockbrains whose guide and captain was Peter, a reader of that Church, watched this woman coming home from some place or other, they pull her out of her chariot: they hail her into the Church called Caesarium: they stripped her stark naked: they raze the skin and rend the flesh of her body with sharp shells, until the breath departed out of her body: they quarter her body: they bring her quarters unto a place called Cinaron and burn them to ashes.'[6]

The woman's name was Hypatia and her death illustrates the historical and psychological elements that are so important to me in the writing of this book. The reader will remember from chapter two that *mortificatio* is the alchemical operation in which the matter being worked upon undergoes a time of putrefac-tion and decay. Thinking of this psychologically as Edinger and Jung did, the ego does not initiate this perilous journey; however, it is inevitable that life cir-cumstances will drag one down into the descent feeling of the *mortificatio*. This process must occur for individuation to progress and precedes *coniunctio* in which the Self will subsequently be encountered. In the story of the alchemist Hypatia, we see how humanity employs *mortificatio* by sacrificing a person in order to avoid the dread that can accompany the birth of new knowledge as well as the patriarchal fear of the feminine. Her story is included with the women alchemists in light of her philosophy and her fervent exploration of the nature of the universe.

Hypatia lived in Alexandria (as did Maria Hebraea, but at a later date) during the 4th century (355 or 370-415 CE). Hypatia's Alexandria was experiencing a point in time when Christianity was struggling to subjugate paganism. Many Christians were obsessed with the notion that the study of mathematics and the work of the natural philosophers was a clear path to Satan.[7] These were the same type of people who insisted the world was flat which was ironic in that the spherical nature of the earth had been recorded for eons, at least as early as humanity was conscious of the planets.

Hypatia is described as one of the few women scientists to be formally acknowledged by historians. Her story begins as so many tales do, years before she actually became the object of that ruthless killing. The entanglement of religion, politics, and social class played a role in her fate; thus, to place her in historical context, we go back to the time in the second century when the Roman Empire had experienced an internal division between its eastern and western domains.[8] The west included nations that we now call Italy, Spain, France, and the United Kingdom. Latin was used as its common language. The eastern empire which would eventually become the Byzantine Empire, clung to its Greek roots. The difference in language only reflected a much deeper philosophical divergence. However, our story focuses on a time when Emperor Constantine had founded Constantinople and had also made the shrewd decision to attempt to unite his people under one religion, Christianity. The Edict of Milan had been issued in 313 CE declaring that it was no longer acceptable to persecute people for following their own religion. Constantine's support paved the way for the Church of Alexandria to become a forceful player in politics as well as religion. St. Mark the Apostle is believed to have been the first leader or Patriarch of the Church of Alexandria.

Constantinople was the gathering place for an assortment of religious leaders from the Church of Alexandria at the Council of Nicaea, May 20-July 25, 325 CE. One of the many issues they discussed and voted upon was the divinity of Christ. The Arian Christians (such as the Visigoths) believed that Christ was *of* God but *not the same* as God. That would make the claim that Christ was an equal in the Holy Trinity problematic. Other believers disagreed and argued that Christ was equal to God. The Nicene Creed was created and has been revised but is mainly the same recitation heard in 21st century modern churches. The Creed made it very clear that Jesus and the Holy Spirit were part of a trinity and that it would be apostasy to preach any other variant. Being saved became a matter of subjugating the old ways that were associated with goddess worship and considered pagan. It is clear that the patriarchy was coming into its full power at the expense of the divine feminine, at least in the Western tradition. It will become evident that Hypatia represented much of what the early church was trying to annihilate.

Hypatia's life has inspired a number of historians but actual records of her work have disappeared. Much of what we know about Hypatia comes from the letters and published work of her students as well as the 5th century Christian historian, Socrates Scholasticus.[9] Hypatia's father, Theon, was a mathematician and astronomer at the magnificent Alexandrian Museum. Theon made sure that Hypatia received an advanced education. Thus, this daughter of much privilege grew up among the writings of Plato, Aristotle, Hermes Trismegistus (the well-known alchemist) and Ptolemy. Eventually, she taught mathematics and philosophy at the Museum and as well, conducted lectures for her myriad students. Hypatia wrote in the area of mathematics, examining the work of Diophantus (algebra), Apollonius (conic sections), Euclid, and Ptolemy. She also developed apparatus for distillation and measuring specific gravity. Although her own work has not survived, scholars believe that Hypatia's writing was probably published as Theon's work, a common practice for centuries to come.

Hypatia's ideas are reflected upon in depth in the correspondence of one of her pupils, Synesius, who eventually became Bishop of Ptolemais and was a faithful friend throughout her life. Although Hypatia's father appears to have been more versed in alchemical writings, her philosophy reflects the alchemical paradigm of an interconnected universe. She was of the Neo-Platonic school, a common philosophy we will see appearing in the work of many alchemists in which knowledge comes from an inner connection to the Divine rather than Divine Revelation, an external experience.

Hypatia embodied the role of teacher. Her students wrote about Hypatia and her teaching with the Eros that comes from the archetypal teacher-student relationship. For example, Synesius uses descriptors such as "blessed lady"[10] and "genuine guide in the mysteries of philosophy."[11] She carried a kind of numinosity for these young male students who again described her as a "divine spirit."[12] Dzielska concludes that Hypatia was actually closer to middle age in the 390s which further enhances the sense of the teacher-elder archetype. She argues that it is clear that Synesius would not have participated in his comprehensive and complex philosophical studies with anyone his own age as teacher. Despite Hypatia's high standing which was most unusual for a woman, her students did not include women. She stayed within the confines of the higher class to which she belonged and as well, appears to have renounced physical intimacy.

Hypatia's theory of existence described people as having the means to individuate within their own psyches as did Plato and later, as developed by Jacob Böehme. Gnostic belief systems come to mind when Dzielska writes of Hypatia teaching that this inner wisdom was "hidden deep inside us waiting for release, [it] renders that individual a bearer of the transcendental world, making him capable of bursting the shackles of matter."[13] People had great potential and only needed a teacher to point them on their way.

Hypatia fell victim to the politics of the times. Theophilus I, another Patriarch of the church (385-412) destroyed a pagan temple and was succeeded by an unsavory fellow, Cyril (412-444). Cyril became the Patriarch of the Alexandrian Church in 412 CE. He is described as being fanatical about his Christian beliefs which when combined with his lust for power, made him a dangerous foe.

He first focused on purging Alexandria of Jews and then turned his focus on the Neo-Platonists whom he considered pagans. Hypatia's influence on both her students and followers was a problem for Cyril as they were from the elite class and often held high positions of authority in both the civil government and the church. Thus, in a manner that would be worthy of modern "spin doctors" in Western politics and advertising, Cyril's supporters were able to convince the mainstream citizens that Hypatia was a witch who practiced black magic, despite the lack of any credible evidence. The stage was set for events to unfold that led to Hypatia's murder, at either the behest or by tacit consent, of Cyril. Hypatia represented too much of a threat to Cyril in light of her popularity and influence among the Alexandrian elite. It might astonish the reader that the Catholic Church eventually canonized this same Cyril. Hypatia was not the first, nor would she be the last person to be killed due to the perception by Ecclesiastics of the threat to their power.

As I mentioned earlier, I have included Hypatia's story in this book although there is no recorded evidence, other than her education, that she practiced alchemy. Her foci were mathematics and astronomy. However, we will see in the following chapters a repeating motif of the masculine principle rising up to quell and destroy the feminine if its power felt threatened by the old practices, religious beliefs, and especially, goddess worship. This will be discussed in greater depth in a subsequent chapter. For now we can be conscious of a pattern of behavior that will be revisited time and time again.

Notes

1. Walter Endrei, *Old Chemical Symbols* (Budapest: Hungarian Academy of Arts and Crafts, 1974).

2. Raphael Patai, *The Jewish Alchemists: A History and Source Book* (Princeton, NJ: Princeton University Press, 1994), 66.

3. Thompson (1932/2002).

4. Margaret Alic, *Hypatia's Heritage: A History of Women in Science from Antiquity Though the Nineteenth Century* (Boston: Beacon, 1986), 37.

5. Ronald D. Gray, *Goethe the Alchemist* (Cambridge: Cambridge University Press, 1954), 150.

6. Alic, 45-46.

7. Alic (1986).

8. Andrew Jotischky and Caroline Hull, *The Penguin Historical Atlas of the Medieval World* (London: Penguin, 2005).

9. Maria Dzielska, *Hypatia of Alexandria* (Cambridge: Harvard University Press, 1995).

10. Dzielska,47.

11. Dzielska, 47.

12. Dzielska, 47.

13. Dzielska, 48.

Chapter 4

Alchemy, Daemons, and Lovers

Willingly I yielded to your prayers, willingly I accepted you / As steel is attracted by the magnet, I follow you myself / Among so many of noble birth that Denmark nourishes / You are the only one to please me in my   innermost heart. [Verses 239-243, Urania Titani, Tycho Brahe][1]

The stories of the three women who are the focus of this chapter present us with accounts that are similar in their poignancy and how they illustrate the importance that many women alchemists placed on relationships as well as their studies. They were highly intelligent, well born women who put their energy into scientific studies that included alchemy. They were also as prone as anyone to choosing mates that resulted in relationships that had both helpful and destructive consequences.

Sophie Brahe: Carrier of Sulphur

Traditional Virgo Traits

Modest and shy
Meticulous and reliable
Practical and diligent
Intelligent and analytical
On the dark side....
Fussy and a worrier
Overcritical and harsh
Perfectionist and conservative[2]

Sophie Brahe was born on August 24, 1559 under the astrological sign, Virgo. This is a fitting place to begin her story considering her work in astrology was lifelong. Sophie Brahe's alchemical work also invokes the image of Venus and the way a relationship with another individual can play out in one's life with *both* positive and negative results. It may seem that the end of her story takes place in 1613 when her second husband, Erik Lange, died. Sophie, first widowed at the age of 29 years, found herself widowed again at the age of 54 years when many women of that century would be considered fairly old and expectations of intellectual achievement would be few. Yet, in 1626 at the age of 67, she published a genealogy of 60 Danish families of the nobility. Sophie lived until the age of 84 years and was buried next to her first husband, Otte Thott, in 1643. On the one hand, Sophie lived an adventuresome life, full of intellectual work that she achieved sometimes at her brother's (Tycho Brahe) laboratory at Uraniborg but often as not, at her own laboratory at Eriksholm. However, at one time Sophie also found herself living in poverty with a husband who had lost both his and her fortunes, chasing the alchemical dream.

Early Biography

Sophie Brahe (1559-1643) was born in Denmark. The Brahe family was very powerful politically, belonging to the ruling group called the Rigsraad, Council of the Realm. This group of families basically ruled Denmark, even to the point of choosing the king.[3] Sophie's parents were Beate Bille and Otte Brahe whose marriage in 1544 was more of a political alliance. Interestingly, married Danish women in the sixteenth century retained their maiden name. Otte died when Sophie was only 11 years old. Beate lived to the age of 78 years.

Sophie's eldest brother, Tycho Brahe, was to become the very famous astronomer. Sophie was eclipsed by her renowned brother and thus, much of the information I have found on her life has been in biographies of Tycho.[4] Sophie was the youngest of eight children and a favorite of Tycho who admired her intellect above that of most women he knew. Tycho liked to refer to himself as Apollo and to Sophie as his muse, giving her the nickname, *Urania*, Goddess of astronomy.[5]

Tycho was actually born a twin; however, his brother died as an infant. He was baptized Tyge after his paternal grandfather but later used the Latin form of his name, Tycho. In a bizarre set of circumstances, despite being Otte's firstborn son, Tycho ended up being raised by his uncle, ostensibly because Uncle Jørgen had no children of his own, especially boys who could carry on his line, whereas, Otte had eight children with four being boys. It would seem Otte could spare his brother an heir.

A brief discussion of Tycho's work provides context for understanding Sophie who assisted him frequently in his alchemical studies as well as with his more widely known work in astronomy. Tycho attended university at Rostock where he met other student-alchemists. John Christianson claims that Tycho was not an alchemist; however he identifies Tycho as an iatrochemist which leads me to believe he is making the common mistake made by some historians as I have pointed out previously, in that iatrochemistry has been associated with Paracelsus and a healing focus, dismissing the connection to alchemy. I wish to reiterate my position that iatrochemistry *is* alchemy despite its focus on healing.

Tycho's schooling was unremarkable being that of a nobleman's son. Sons of nobility attended the university taking specific courses, but did not necessarily register to obtain a degree as they did not need one to further their already guaranteed place in society. Tycho had a typical education for his time including: logic, debate, rhetoric, and Latin but he was also captivated by astronomy, arithmetic, geometry and music (the *quadrivium*).

According to Victor Thoren, Tycho was conversant with alchemy and concurred with the ideas put forth by Paracelsus, that everything in the universe was connected with some type of chemical glue. Eventually, he constructed alchemical labs in his basement as well as in his dining room at Uraniborg on the island of Hven. The island was a gift from Frederick II for the construction of an observatory and other buildings Tycho would need to further his science. Uraniborg translates to "Heavenly Castle"[6] and was a functioning community with workshops, windmill, paper mill and whatever else was needed to support Tycho's work. It became quite famous and for example, on March 14, 1590, King James VI of Scotland made a visit. In due course, Tycho added a print shop to Uraniborg, which made printing his work much easier. Unfortunately, there are no surviving papers from Tycho's alchemical practice as opposed to his voluminous publications in astronomy.

Tycho was mentored in the study of alchemy by Thaddeus Hayek, a physician from Bohemia who claimed that he had witnessed John Dee, another well-known alchemist, produce gold from mercury in 1584. Furthermore, Tycho read Petrus Severinus's work on Paracelsus that outlined alchemical practice. Tycho referred to alchemy as "terrestrial astronomy"[7] and wrote that the study of alchemy had led him to "'a great many findings with regard to metals and minerals as well as precious stones and plants and other similar substances.'"[8] The term terrestrial astronomy is significant in that alchemists believed the working of the cosmos could be discovered in the science of transmutation – *as above, so below*. Thoren speculates that Tycho was more dedicated to the Paracelsian notion that the true goal of alchemy was to create healing elixirs. On a list of visitors to Tycho's lab on Hven, a fellow is listed frequently in the years 1590-1597 as, "Paul, the pharmicist."[9] Thoren argues: "Next to astronomy, alchemy was Tycho's greatest intellectual passion."[10]

Tycho is more well known for his work in astronomy having observed a supernova in 1572, adding to the mounting evidence that celestial spheres were not static but indeed, subject to change. This thinking was in agreement with that of Copernicus and would be a notion that would land Galileo in trouble with the Catholic Church. By 1574, Tycho had published *De Stella Nova*, about the star he discovered. Tycho also played a part in Johannes Kepler's work on planetary motion and made observations of the Great Comet of 1577.

Tycho was known as a demanding employer and researcher. At some time he appears to have lost control of his temper as his nose was partially severed in a duel. He wore a prosthesis made of copper, gold and silver. Tycho married a commoner, Kristine, with whom he had five children but despite the seeming romance of marrying a woman who was not of noble birth, it is believed he did so more out of convenience. His true passion was his work and Sophie was often a co-worker in his endeavors.

As with so many women in this book, Sophie received her education at home, with Tycho sometimes acting as tutor in subjects such as chemistry. She was well versed in German, Latin, astronomy, alchemy and classical literature. Sophie was a skillful poet and she enjoyed casting horoscopes for friends. Astrology was considered a legitimate science by alchemists as well as other natural philosophers. Sophie was a quick study and in 1573, when she was 14-years-old, she assisted Tycho in observing a lunar eclipse. Christianson has included a lovely quote from Tycho's writing concerning Sophie.

> But I seriously warned her to desist from astrological speculations, because she should not strive after subjects too abstract and complicated for a woman's talents. But she, who has an unbendable will and such great self-confidence that she will never yield to men in intellectual matters, cast herself all the more energetically into her studies and in a short time learned the basic principles of astrology, partly from Latin authors whom she had translated into Danish at her own expense, and partly from German authors on the subject (for she has an excellent grasp of that language). When I saw the clear signs of this, I quit opposing it and simply advised her to moderation in her ongoing studies.[11]

Tycho also wrote a letter comparing Sophie to *Fulvia Olympia Morata*—a famous Italian scholar. Sophie is described as strong-willed and not to be cowed by at least, Tycho's sense of what was proper for women and what they were capable of doing. He clearly underestimated women but again, was a man of his times.

Sophie must have been quite the firebrand, which is why I have referred to her as a carrier of sulphur, the heat or fire symbol in alchemy. Sophie's alchemy focused on both the pursuit of transmutation of base metals into gold as well as the creation of numerous medicinal cures. She developed a "pest-elixir"[12] that was supposed to combat the plague.

Sophie married Otto Thott of Eriksholm at the age of 19 or 20 and gave birth to one son, Tage on May 27, 1580. Otto died in 1588 and there is some question whether she stayed on and managed Ericksholm or returned to her family home.[13] Nevertheless, Sophie, a young widow, plunged into her studies of horticulture, astrology, and Paracelsian medicine and as well, built a laboratory in her garden. Sophie made frequent visits to Tycho on his island. Uraniborg gave her a place to not only work with her brother but to pursue her own studies in astrology and alchemy.

While working with Tycho on his island observatory in 1590, Sophie met Erik Lange, who is described as a romantic figure, having escaped the massacre of Protestants in Paris in 1572 (the St. Bartholomew's Day massacre) "by hiding in Parisian back alleys and fleeing through the sewers."[14] Erik studied at Tübingen University in Baden-Württemberg, Germany as well as the University of Leipzig so was not a dilettante, although he would go on to act quite foolish when it came to alchemical studies. Lange's sister, Margaret, had married one of Tycho's brothers, Knud, and Erik visited Tycho frequently to discuss his own work in alchemy. Tycho and Erik became close friends and perhaps it was inevitable that Sophie and Erik fell in love and became engaged. By 1590 Sophie and Erik were betrothed but within a few months, Erik was under house arrest for a brief period for unpaid debts. Tycho was the one member of the Brahe family who did not oppose their relationship completely; however, in 1594 he composed a poem titled, "Urania to Titan" which is in the form of a letter from Sophie to Erik, trying to talk some sense into him. The details that gave the family cause to be anxious about Erik surrounded his alchemical work.

Upon the death of Erik's father, he had inherited the estate, Engelsholm, with his sister, which provided him with the means to plunge fully into his consuming interest in alchemy. Erik sold a portion of his estate to obtain a title and the position as governor at Bygholm Castle. He then proceeded to lose his fortune on failed alchemical experiments. Sophie and Erik remained engaged for about 12 years despite the loss of his castle, his fortune, and his need to flee Denmark to escape creditors. Erik did not appear to be stable husband material; however, Sophie was in love and that can explain much.

It seems that Erik Lange was caught up in a web of obsession regarding alchemy. He was certain that alchemical processes were vital to the core of our existence and could be utilized for one's own good. He was desperate to prove his theories and would do anything he could to achieve his goal. Lange apparently believed, for example, that he had been able to multiply a quantity of flour using alchemical operations, a useful and practical achievement according to Tycho.

After many years, Sophie was also in debt and we can surmise that she had been giving money to Erik. Sophie appears to have been in Hamburg by the summer of 1601 where she was staying with a physician and his family while he was treating her for melancholy. Tycho moved to Prague and wanted Sophie to join him but he died on October 21, 1601 before she could decide whether or not to join him. Sophie and Erik did marry on March 21, 1602, a few months after Tycho's death. One might speculate that Tycho's death was a huge blow for Sophie. He had been both brother and father to 11-year-old Sophie when their father died. He was the older brother who was smart and heroic but also, one who acknowledged Sophie's talents and intellect. It appears that Erik was about 20 years her senior therefore by this time, Sophie was in her 30s and he was middle aged, in his 50s. It may sound a cliché but considering that the age difference between Sophie and Tycho was 13 years, it is likely that Erik may have been an unconscious substitute for Tycho. Marriage to Erik may have been inevitable.

The couple was, and continued to be, in grave debt. Sophie's sisters gave her jewelry and clothes that she pawned and was never able to recover. But our indefatigable Sophie turned her grief and energy to her studies and to her work in chemistry.

Subsequently, Erik died impoverished in 1613. Sophie's son by Otto, Tage, who became very wealthy from the inheritance of the Thott family estate, did provide a home for his mother until she died in 1643. In addition to chemistry, Sophie had a deep interest in researching genealogies. In 1626 she had completed work on 60 Danish noble families. The story is told that Sophie taught her servant at that time, Live Larsdatter, the healing arts. It was claimed that Live lived until she was 123 years old. Sophie is buried with the family at Eriksholm next to her first husband and Tyge's father, Otto.

What Was She Thinking?

How many people have been asked that question by someone looking back at what seems like poor decision-making? Reading through the lens of most historians, including myself, we can imagine that Sophie had a hard and bitter life with Erik. Christianson mentions her humiliation and tears; I wonder why she stayed with him. These are the kinds of interpretations that may be accurate or again, might not.

The historical imagination asks us to consider different interpretations. As a dear friend pointed out to me, might Sophie have been happy with Erik because she just loved him? Is it possible that she was not relieved, as I imagined, when he died. Perhaps she threw herself into her work and completed the 900-page genealogy as a means for channeling her grief. Without her letters or diary, or other evidence, we can make a logical guess, knowing that we see this thing through or own lens, our personal filter.

Sophie and Erik's story also supports the notion that alchemy was about many things to different people. Depth psychologists like Carl Jung, write about the male aspect of a woman's psyche, the *animus*. Women are often deeply attracted to men who carry or embody their *animus*, which may or may not result in a successful relationship.[15] As we learned in chapter two, one of the goals of healthy psychic development is to recognize and become comfortable with these aspects of ourselves.

It is clear that Sophie was every bit as able to attain the intellectual achievements as her brother; however, cultural mores regarding women's advancement in academic fields narrowed her options. Erik, however, could go where she could not and thus she found the perfect alchemical partner. Erik might be perceived as a failed alchemist but he was *one of the in-group* that Sophie could only watch from the margins. They strove to create gold and healing elixirs and as well, were often admired by society. At the same time, both Erik and Sophie became so identified with the work that they lost sight of the need to stay grounded in daily life. As has been the story for many alchemists, they spiraled into poverty and never realized their dream.

It is tempting to simplify Sophie's devotion to Erik and reduce it to something like a compensation for her lack of a father and her devotion to Tycho who had died recently. Yet, invoking Hillman's Acorn Theory causes me to wonder if Sophie was able to more truly perceive the essence that was Erik.[16] Hillman examines the core nature of a person and how that core, or acorn, can be overlooked amidst the layers accumulated during one's life. A biographer tends to analyze and look at the subject's childhood and significant life events, drawing a conclusion regarding the adult. What may be missed is that the acorn has a hidden but profound effect on one's development. Hillman illustrates this by discussing Henry Kissinger who was said to state that his childhood exposure to Nazism was not as overwhelming as one might think. Biographers concluded otherwise, but Hillman suggests that the acorn that was the core of Henry Kissinger at the deepest level of his psyche may have been more integrated than what has appeared to psychologists to be a compensation for childhood trauma. A woman can appear to study and write as a means to compensate for the girl who did not feel special and needed to be seen as exceptional by her accomplishments. That may be part of her story but the acorn of the woman is the scientist. She needs to study, learn as much as possible, write it down, share it and teach as if her very existence is linked to this work.

Sophie's keen intelligence might also have allowed her to tease out the layers of what made Erik special and the man she loved so dearly. It is difficult to formulate a definitive conclusion in the absence of Sophie's words. It does caution us to take care in pathologizing women who are attracted to seemingly unsuitable men. This is not to say that a person who is in relationship with an abusive partner should stay in that situation. It just suggests that before concluding that someone's obsession with someone else is nothing more than a simple result of this or that in his or her past, a closer look and a deeper inspection for what it means needs to be taken. The act of peering closer at the parts that make up the whole requires a type of *separatio* like that discussed in chapter two. The different aspects of a person should be considered before reuniting them to ascertain a sense of the complete person. The whole is usually greater than the sum of the parts.

Anna Maria Zieglerin and Her Daemon

Imagine a young woman about whom it was told, had survived a premature birth by being wrapped in an unusual blanket consisting of some unfortunate deceased woman donor's skin for a period of 12 weeks.[17] This bizarre incubation technique was thought to have allowed her to maturate into the viable baby we know as Anna Maria Zieglerin (c. 1550-1575). Tara Nummedal has written an account of Anna Maria's story that serves as an illuminating example of how so often women alchemists bridged the space between spiritual and physical alchemy. Anna Maria's family was of modest German nobility. In addition to her remarkable birth story, Anna Maria had not begun menstruating, even by her early teen years, causing her to believe she carried a special kind of purity. Tragically, she was raped by a potential suitor, Nikolaus von Hamdorff, when she was 14-years-old and became pregnant, which possibly explains why despite her supposed absence of a menstrual cycle, she carried a virginal self-image of being other-worldly, pure, and chaste. The baby was born at term but was allegedly drowned by Anna Maria and her nurse.

Anna Maria's first marriage was a disaster. At the age of 16 years and after only nine weeks of marriage, she was widowed from her husband who died in a riding accident. She was then coerced into marrying a very unpleasant fellow, Heinrich Schombach, who is described as a "cross-eyed court jester."[18] Heinrich must have been a less than charming suitor as he was also known as "Harry the Squint."[19] The marriage was not a congenial one.

In the course of their chaotic relationship, the couple met an alchemist, Philipp Sömmering, for whom Heinrich became an assistant. By 1571, Anna Maria was 21-years-old and accompanied her husband to join Sömmering in his alchemical work for Duke Julius of Braunschweig-Wolfenbüttel. Their task was to create the Philosopher's Stone in order to subsequently create gold and gemstones for the Duke's enrichment.

There are some discrepancies in the stories surrounding Anna Maria's true motives for her alchemical work; however, Sömmering was a criminal and saw this scheme as a way to make money in the tradition of many alchemical swindlers. He found the ideal accomplice in Anna Maria's husband, Schombach.

Despite the well-known stories that fakers or puffers would try to swindle gullible people who were obsessed with obtaining gold by any means possible, it was not uncommon for members of the aristocracy, seemingly well educated, to fall prey to the lure of easy riches. Heinrich Schombach was a willing participant in this plan along with another fellow villain, Sylvester Schulfermann of Lübeck. Schulfermann was later accused of the suspicious murder of a courier who had been about to deliver information to the Duke that would have been damaging to Sömmering's plans. Yet, amidst all this scheming, it appears that in the beginning of this intrigue, Sömmering truly believed that Anna Maria knew how to produce the Philosopher's Stone. After many failed attempts, he concluded that no stone was going to appear. He became anxious considering the potential consequences of being found out to be a fraud by his benefactor and requested to be released from the contract along with his colleagues, Anna Maria and Harry Schombach. However, when the Duke asked for repayment of the 2000 talers he had already paid them, Sömmering fled, incriminating the group and leading to their arrest.

The threesome found themselves on trial in 1574 for a number of crimes including: "murder of a courier, attempted poisoning of Duchess Hedwig, and copying keys to the Duke's chambers with intent to steal some papers."[20] However, their real crime seems to have been the inability to produce a genuine Philosopher's Stone, as well as the subsequent attempt at a cover-up. Nummedal notes that confessions to all charges were tortured out of the three defendants. They were put to death on February 7, 1575. Sömmering and Schombach were burned with tongs repeatedly and quartered. Anna Maria was burned alive, "strapped to an iron stool."[21] She was 25-years-old.

If anything helpful can come out of reflecting upon Anna Maria's horrific experience, it is that examining the records of her interrogation, miscellaneous papers in which her work was documented, and testimony from her collaborators gives us a glimpse of her own alchemical investigations. Tara Nummedal's archival research is an important example of the need to look further than traditional historical sources to fill in the blank spaces in the history of alchemy. Anna Maria was not merely an assistant to her male colleagues as some historians of alchemy have portrayed women alchemists. She seems to have truly believed in her ability to create the Philosopher's Stone. She wrote a paper in 1573 titled, "Concerning the Noble and Precious Art of Alchamia"[22] and ran her own laboratory with the help of a personal assistant. Further evidence shows that Anna Maria once wrote a letter to Duke Julius in 1573 in which she enclosed a stone that she claimed was a genuine Philosopher's Stone.

> 'I am sending you this small lump; the greatest little stone [i.e., the Philosopher's Stone] I have set again in the wine so that it does not entirely dissolve into the air . . . in a short time I want to show Your Princely Grace something greater . . . Tonight with the help of God in heaven we want to begin the two pounds of quicksilver.'[23]

Anna Maria's alchemy was a mixture of the odd and the familiar alchemical practices. For example, one of the operations Anna Maria employed was feeding what she called, philosophical oil, to a bird, prior to baking the bird for later use. Birds are common alchemical symbols.[24] Typically, birds represent the vaporization of the elements being worked upon. One sees various species represented but a common one, the pelican, is often shown pecking her breast to provide blood to nourish her young. The alchemical vessel named after the pelican resembles it and the process is a metaphor for the work.

The oils Anna Maria produced had myriad uses in her alchemical work. Anna Maria believed that the final creation of the Philosopher's Stone required that two stones, to be created in the course of her work, had to be combined in the final alchemical operation of *coniunctio*. Anna Maria's Philosopher's Stone was said to be ruby-like and "a mere half of this stone, Anna Maria maintained, could tinge sixty pounds of lead into gold; the other half could be multiplied for future use." [25]

Anna Maria also claimed that she could produce gems such as diamonds and sapphires, use her alchemical oil to help produce ripe fruit out of season, cure myriad diseases and furthermore, assist in the conception of a child. Upon birth, the alchemical child could be raised to at least adolescence, by drinking her oil exclusively. In the tradition of most legitimate male alchemists, Anna Maria labored away while believing that she was infused with the spirit of God. Genuine alchemists believed that God must bless their work for it to be successful. They often began the *opus* with prayer and refused to share their work with those they considered to be unworthy.

One can only wonder at the psychological confusion of this young woman who lived with both a vast intellect as well as the troubled psyche of someone who had experienced a rather unusual birth story, survived a rape, allegedly murdered the resulting child and undergone widowhood in her first 21 years. Anna Maria's alchemy was surely affected by these experiences which yield an odd mixture of science and superstition. Yet, in a departure from the traditional norm, Nummedal points out that Anna Maria believed that women were critical participants in alchemical work. Many alchemists had attempted to create the homunculus, a tiny human being grown in the alchemical vessel called the alembic. Anna Maria was alone in that she believed that in so doing, the homunculus needed to grow in a woman's womb, as opposed to maturing in a mere earthen vessel. The philosophical child would be a product of sexual union and the use of her alchemical oil, "Lion's Blood,"[26] would ensure viability and growth. The feminine was seen as equally critical in the work that was mostly dominated by the patriarchy.

The Count

A *daemon* is a term we encounter in archetypal psychology. It is a word from Greek mythology that refers to a kind of deity or spectre. *Daemons* can be messengers between the gods and humankind. They can also appear to some cultures as angels as well as being mistaken for demons. Nummedal relates a story about Anna Maria that I believe illustrates the idea of a *Daemon* clearly. This helps us see into the psyche of one of the women alchemists and how she related to her work for I am not merely interested in learning the identities of women alchemists; I want to understand why they studied alchemy. Anna Maria presented herself as a participant in a story that is considered to be fictitious regarding the existence of a Count Carl von Oettingen, whom she described as an illegitimate son of the noted alchemist and physician, Paracelsus. Most of this story was gleaned from Anna Maria's trial testimony as well as that of her co-conspirators who were familiar with her tales of the Count.

Count Carl was described by Anna Maria as her alchemical mentor. Count Carl had supposedly given Anna Maria a powder that could turn lead into gold. He taught her the art of alchemy while she was living with her mother. Anna Maria believed that she and Count Carl were destined to be the parents of several children via some alchemical process that is unclear to the reader. They would be able to produce one child per month, who would then be raised by the Count, with Anna Maria's elixir, Lion's Blood oil. Lion's Blood usually refers to the dissolving of red sulphur in mercury to create the Philosophers' Stone.[27]

The children would be girls, who like their mother, would not menstruate and would have the potential to live to be as old as Methuselah. But Anna Maria reminds the reader that in reality, the girls would not live quite that long since the sin of Adam denied humankind its ability to live a very long life. These girls would have created the foundation of a better world. We will see that this same yearning for a utopia will also be reflected in the work of many of the seventeenth century women alchemists but in the latter cases, arose from a more traditional, theological framework than Anna Maria's.

Nummedal argues that Count Carl and his *existence* allowed Anna Maria to claim some legitimacy as an alchemist in that she did not claim the knowledge as her own creation, but as the teachings of Count Carl and thus, Paracelsus, his father, a truly legitimate alchemist. It should be understood clearly that Anna Maria did not present Count Carl as a simple fiction. She relates a complex background for him that she was able to describe in detail. Count Carl's mother may have had illegitimate relations with Paracelsus but she was the vehicle for Count Carl receiving a book that held the secrets of Paracelsus' alchemy. In a rather Messianic sounding story, Anna Maria relates that Count Carl's mother was supposed to have predicted that she carried a child who would be a greater alchemist than Paracelsus as well as all other alchemists who preceded him. Continuing the Christ image, Count Carl's birth was foreshadowed by a sign.

'A white eagle fell out of the heavens,' Sömmering [Anna Maria's partner] re-counted, and 'when she [the Countess] went riding, she saw the eagle laying there on the ground, and then it raised itself again up to the heavens.'[28]

The white eagle is another alchemical image that is associated with philosophical Mercury, not necessarily the actual element. As Mercury is a messenger, the eagle represents the spirit of the matter rising from the *prima materia*, *sublimatio*.[29] The color white was named throughout the alchemical process. It referred to the calcinatio phase of burning the substance to ash. The color white also takes on significance as a phase called *albedo*.

The second whitening or *albedo* was conceived as a state of illumination or the dawning of the unknown personality in consciousness. Some [alchemists] deemed the albedo the attainment of the goal. Others believed the opus reached fulfillment only when dawn turned into the ruby brilliance of sunrise.[30]

Paracelsus writes of the need to combine the white eagle or Mercury with the blood of the lion, which may be associated with sulphur. The symbolism gets a little convoluted to read but to simplify, the alchemical operation is one in which elements are combined causing them to transform in a way that leads into the next stage of the process.[31]

The Count was described as a faithful suitor despite Anna Maria's marriage to Harry Schombach. She related that he had turned down his own opportunity for marriage with Queen Elizabeth I to remain loyal to Anna Maria. Interestingly, Harry was fully aware of the Count. Count Carl offered Anna Maria's husband his own sister, as well as a goodly amount of gold, for Anna Maria's hand in marriage. Anna Maria's husband was agreeable, assuming the gold would be placed directly into his hands. Much to Harry the Squint's regret, the gold was not forthcoming.

Anna Maria also possessed love letters, written presumably by her, from the Count that proved his existence and his love for her. Nummedal notes:

> Unlike her male colleagues who cited brief encounters with mysterious adepts, Anna Maria intertwined her studies with Carl and her personal relationship with him. This was an alchemical partnership, forged by destiny and cemented in marriage and family, which carried the potential for no less than the salvation of humanity.[32]

Anna Maria's story of Count Carl can be seen as a *coniunctio* symbol or image that is often portrayed in alchemical engravings depicting a King and Queen united in a royal marriage. Nummedal's discussion of the Count presumes that he is a fictional character; however, I believe that Anna Maria was convinced of the truth of his existence. At the very least, it seems that Anna Maria believed in the idea of the Count, a positive masculine figure with whom she could unite and fulfill an idealized destiny. Her unconscious motive may have been similar to Sophie's Brahe's attraction to the masculine embodiment of her own dreams in Erik Lange.

Nummedal's further contribution to the women alchemists' stories is her observation that for some women, alchemy focused on the affirmation of the reproductive cycle and its connection to a larger purpose. Regarding Anna Maria's motivation she writes, "Imagining herself a new Virgin Mary untainted by the sins of Eve, Anna Maria hoped that alchemy ultimately would help her to repopulate a fallen world."[33] When I hear women's stories such as Anna Maria's, it is clear that their alchemy retained the idea that the natural world was inextricably tied to the spiritual domain. Although it is clear that there were women who participated in alchemy in order to hoodwink a wealthy victim, most of the stories I have encountered involve women who saw alchemy as part of a larger purpose. Furthermore, they were not quite ready to accept that the world was a soulless machine. We will see in subsequent chapters how seventeenth century women alchemists seemed to struggle with the increasingly uneasy relationship between spirit and matter.

Susanne Katharina Von Klettenberg: Goethe's Teacher and *Maggid*

> *Skirting all rational barriers which a new Enlightenment had put up against the mystical quest for the philosophers' stone, a tiny rivulet of alchemic thought had worked its way into German classicism.*[34]

The Kabbalah teaches that when the student is ready, the teacher, or *maggid* will appear to the seeker. The *maggid* may be someone the seeker already knows, may be an unexpected stranger, or may even be a Divine spirit. He or she may offer words of encouragement but most likely just has words that help the seeker find the answer within her own psyche. The *maggid* says the right thing at the right time, recommends the helpful poem or book, and tells the needed story. Yet, the *maggid* also relies on guidance from the Divine, creating a sacred circle between teacher, student, and God.[35]

Johann Wolfgang von Goethe (1749-1832) spent the winter of his 19th year in Frankfurt (1768/69). He needed to recuperate from a bout of bad luck including an illness, a broken romance, and the shock of the murder of a German archeologist and art historian he admired, Johann Joachim Winckelmann. Goethe's physician, a Dr. Metz, is described as one "who recommended certain mystic, chemical-alchemical books to those of his patients who showed any inclination in this direction."[36] Goethe was hesitant and skeptical concerning these non-rational ideas about medicine but was encouraged to consider them seriously by Susanne Katharina von Klettenberg, a friend of his mother's. Susanne Katharina believed, "that body and soul could only be cured as a unit." [37] In order to heal, one had to treat the body and soul holistically. She had secretly studies alchemical writings and hoped to persuade Goethe to consider thinking about his health in a different way. She became both a teacher and the embodiment of positive feminine energy that Goethe ultimately incorporated in his writing.

Susanne Katharina von Klettenberg was both friend and part of the same religious sect as Goethe's mother. By the time Susanne Katharina met Goethe, she was in her mid-40s and remained unmarried. Her uncle, Johann von Klettenberg, had been beheaded before her birth and appears to have had occult interests that he passed on to his niece. Susanne Katharina is described as being very devout. She worshiped as a Pietist, a sect associated with the philosophies of Jacob Böehme, called the Herrnhuter.[38] Böehme employed alchemical symbolism in great detail in his writing such as that found in the 1621 treatise, *De signatura rerum*.[39] The nexus of alchemy and Christianity was evident in Susanne's diligent study of alchemy. This may seem unlikely to the modern reader who may think of Christian theology and occult studies to be at odds with each other. When we remember that the more serious alchemists were also very devout and hoped to understand the nature of God, this makes more sense. The devoted alchemist believed that his work would *only* succeed if it was blessed by God and *that* would only happen if he was pure in heart. Susanne's consuming piety and asceticism are described by Ronald Gray.

> Once interpretation begins, one is confronted with the vast and often contradictory variety of behaviour which alchemy and occultism inspire in their adherents. For Fräulein von Klettenberg, the process of 'Nigredo', the descent into Hell, meant the complete renunciation of earthly pleasures, of love and dancing and fine clothes, and the annihilation of self-will. In other Pietists who derived their beliefs from the same source, but who were less ascetically or mystically inclined, it took the form of acute repentance and contrition.[40]

Despite the differences in their ages, Susanne Katharina seemed to have a calming influence on Goethe. Despite Goethe's initial reluctance toward alchemy, while suffering from a particularly uncomfortable bout of some digestive illness, his mother convinced him to give Dr. Metz's special elixir a try. Upon experiencing significant relief from the symptoms, Goethe began paying closer attention to the alchemical work practiced by Susanne. Gray quotes another scholar, Lavater, who described Susanne Katharina: "'she has an incomparable insight into chemistry, the nature of colours, etc.'"[41] She became Goethe's teacher of alchemy. He wrote in his autobiography:

> 'My friend,' . . . 'who lived alone as an orphaned only child in a large, well-appointed house, had started some time past to acquire a small blast furnace, as well as alembics and retorts of modest size. Guided by Welling's [wrote *Opus magocabbalisticum*] pointers and some plain hints from the physician and master [Dr. Metz], she began to concentrate on iron, which allegedly embodied the most curative of powers if one knew how to unlock them. Since volatile salts played an important role in all instructions we read, all these operations required alkalies that, in evaporating into the air, would combine with the supernatural and in the end produce an arcane, but efficacious, saliferous substance.'[42]

Goethe read Paracelsus, Basil Valentine, John Baptiste Van Helmont, George Starkey, and the alchemical treatise, *Aurea Catena Homeri*. Yet, he was not naive as to how others might perceive his interests. Thus, he avoided leaving himself open to hurtful comments as evidenced in an observation he wrote: "'But most of all, I hid from Herder [a friend] my mystical-cabbalistical chemistry and all the things related to it.'"[43] Goethe furnished himself with his own laboratory and began to experiment with the alchemical operations writing:

> For a long time I was primarily working with the so-called Liquor silicum (silicic acid) which is produced by smelting pure quartzite with liberal amounts of alkali. The result is a transparent glass which melts in the presence of air, leaving a beautifully lucid liquid. He who has done this and seen it with his own eyes will not mock those who believe in a virginal earth and the possibility of acting on and through it, now and always.[44]

Susanne Katharina was well versed in alchemical skills but the next section shows how her story is very similar to Anna Maria Zieglerin's.

Susanne Katharina's Great Love

Susanne Katharina described a vision that she experienced in which she underwent a "mystical Union"[45] with Christ. Her encounter sounds very much as if she interacted with a *Daemon* and experienced the sacred *numinosum*. In her vision, Christ appeared to Susanne with such reality that she believed she was able to kiss the wounds on his body as well as experience something close to a physical union. In correspondence she had with J. K. Lavater, she wrote about the experience as something that had a perfecting effect on her.[46] She invokes the term *aurum potabile*, an alchemical medical remedy and whose use is consistent with the beliefs of Jacob Böehme.

The vision is similar to Anna Maria Zieglerin's *Count Carl.* Both women were able to experience a sense of *coniunctio,* or a merged relationship with a representation of a male suitor who embodied all that was good in the world. Both the Count and Christ figures represent the best aspects of the positive masculine principle. The physical practice of alchemy also provided a concretization of the union between the feminine and masculine principles which had a *numinous* or holy aspect. The spiritual aspects of alchemy were inseparable from the physical work in the laboratory. Both women carried the virgin archetype either by remaining celibate or as in Anna Maria's story, by believing she was fundamentally pure. Susanne Katharina embodied the archetype of teacher as well as this kind of virgin-goddess that captured Goethe so strongly. Gray argues that Susanne Katharina was the model for Goethe's Amazon. He notes that Goethe had written *Lehrjahre*, in which he based a character's life on Susanne's. Natalie is the niece of the "'schöne Seele'" who "is portrayed as a woman of great perfection, one who lives a life of quiet devotion and pious harmony."[47] Goethe linked the schöne Seele character to his beliefs about what an Amazon might be like whom he describes as "'a true man-woman, . . . a virgin, a virago in the best sense of the word, whom we can admire and honour, without being precisely attracted to her.'"[48] The man-woman or hermaphrodite is a common alchemical symbol. One often sees it portrayed in old engravings such as that found in the *Rosarium Philosophorum* c. 1550.

Figure 4.1: Hermaphrodite from the *Rosarium Philosophorum* c. 1550

Susanne Katharina van Klettenberg was a spiritual and scientific guide for Goethe which illustrates again how alchemy often surfaced in places where it was not named specifically as such, but provided the theoretical framework for the thinking and world view of its practitioners. In his book, *Scientific Studies*, Goethe writes:

> The things which enter our consciousness are vast in number, and their rela-
> tions—to the extent the mind can grasp them—are extraordinarily complex.
> Minds with the inner power to grow will begin to establish an order so that
> knowledge becomes easier; they will begin to satisfy themselves by finding co-
> herence and connection."[49]

The three women alchemists I have discussed in this chapter embody the opposing aspects of alchemy, physical and spiritual, that would eventually become split off from each other. The next chapter introduces some remarkable women who bridged that ever-increasing gap between religious and scientific practice that would take firm root in the seventeenth century.

Notes

1. *Urania Titani* was written by Sophie's brother, Tycho, and is a letter to her love. This translation can be found in part in Minna Skafte Jensen, *Friendship and Poetry: Studies in Danish Neo-Latin Literature*, edited by Marianne Pade, Karen Skovgaard-Petersen, and Peter Seeberg. (Copenhagen: Museum Tusculanum Press, 2004), 181.

2. *http://www.astrology-online.com/virgo.html*

3. Marilyn Ogilvie and Joy Harvey , eds.,. *The Biographical Dictionary of Women in Science,* volume 1 (New York: Routldege,2000), 170.

4. John Robert Christianson, *On Tycho's Island: Tycho Brahe and His Assistants, 1570–1601* (Cambridge: Cambridge University Press, 2000); John Allyne Gade, *The Life and Times of Tycho Brahe* (Princeton: Princeton University Press, 1947); Charles C. Gillispie, ed., *Dictionary of Scientific Biography,* Volume II (New York: Charles Scribner's Sons, 1973); Victor E. Thoren, *The Lord of Uraniborg: A Biography of Tycho Brahe* (Cambridge: Cambridge University Press, 1990).

5. Christianson (2000); Gade (1947).

6. Thoren, 404.

7. Thoren, 211.

8. Thoren, 211.

9. Thoren, 206.

10. Thoren, 210.

11. Christianson, 260.

12. Gade, 89.

13. Marilyn Ogilvie and Joy Harvey, eds., *The Biographical Dictionary of Women in Science,* Volume 1 (New York: Routledge, 2000), 170.

14. Christianson, 260.

15. Jung also discusses the *anima* or the female principle in the male psyche.

16. James Hillman, *The Soul's Code* (New York: Time Warner, 1996).

17. Tara E. Nummedal, "Alchemical Reproduction and the Career of Anna Maria Zieglerin, *Ambix*, 48 (July 2001): 55-68.

18. Nummedal, 59.

19. Reinhard Federmann, *The Royal Art of Alchemy,* trans. R. H.Weber (Philadelphia: Chilton, 1964/1969), 171.

20. Nummedal, 56.

21. Nummedal, 56.

22. Nummedal, 56.

23. Nummedal, 59.

24. Lyndy Abraham, *A Dictionary of Alchemical Imagery* (Cambridge: Cambridge University Press, 1998), 23-26.

25. Nummedal, 60.

26. Nummedal, 60.

27. See Martinus Rulandus, *A Lexicon of Alchemy or Alchemical Dictionary* (Frankfurt, Zachariah Palthenus, Bookseller, 1612). Also Lyndy Abraham, *A Dictionary of Alchemical Imagery* (Cambridge: Cambridge University Press, 1998), 166-167 defines red lion, also known as red sulphur as "the hot male seed of metals." It was combined with mercury, the feminine principle in a "chemical wedding."

28. Nummedal, 63.

29. See Ami Ronnberg and Kathleen Martin, eds., *The Book of Symbols* (Cologne, Germany: Taschen, 2010), 256-258.

30. Ronnberg and Martin, 660.

31. See Abraham, 64-65 and 166-167 for a more complete exegesis on this part of the alchemical process.

32. Nummedal, 65.

33. Nummedal, 58.

34. Federmann, 225.

35. Halevi (1976).

36. Federmann, 229.

37. Federmann, 229.

38. Gray (1954).

39. Thorndike, 183.

40. Gray, 252.

41. Gray, 101.

42. Federmann, 231.

43. Gray, 54.

44. Federmann, 232.

45. Gray, 22.

46. See Gray, page 22 and H. Funck, *Die schöne Selle*, page 282 where he quotes Susanne's letter to Lavater from 12 September 1774.

47. Gray, 228.

48. Gray, 229.

49. Johann Goethe, *Scientific Studies* (New York: Suhrkamp, 1998), 9.

Alathea Talbot, *Natura Exenterata*, frontispiece. By permission of The Huntington Library, San Marino, California.

NATURA EXENTERATA:

OR
NATURE UNBOWELLED

By the most

Exquisite Anatomizers of Her.

Wherein are contained,

Her choicest S E C R E T S digested into
R E C E I P T S, fitted for the Cure of all sorts
of Infirmities, whether Internal or External,
Acute or Chronical, that are In-
cident to the Body of Man.

Collected and preserved by several Persons of Quality and great Experience in the Art of Medicine,
whose names are prefixed to the Book.

Containing in the whole, One thousand seven
hundred and twenty.

Very necessary for such as regard their Owne
Health, or that of their friends.

Valetudinem tuam cura diligenter.

Whereunto are annexed,

Many Rare, hitherto un-imparted Inventions, for
Gentlemen, Ladies and others, in the Recre-
ations of their different Imployments.

With an exact Alphabetical Table referring to the several
Diseases, and their proper Cures.

London, Printed for, and are to be sold by *H. Twiford* at his shop in
Vine Court Middle Temple, *G. Bedell* at the Middel Temple
Gate Fleetstreet, and *N. Ekins* at the Gun neer the
West-end of S. *Pauls* Church, 1655.

Natura Exenterata, title page. By permission of The Huntington Library, San Marino, California.

Chymicall Extractions.

TAke I.S. ſeven times rectified, and put it in a pelican, and let it Circulate forty dayes; and this is his burning water or quinteſſence or Heaven, and hath a ſmell ſo odoriferous that it being open in any Roome, all that paſſe by will be forced to come in, if this ſmell be wanting when you open it, put it in a bolts head well ſtopped with wax, and bury it in Horſe dung with the bottom upwards, ſome part being bare to the Ayre, and having ſtood ſo many dayes, take it out, and with a ſmall warm wyer bore a hole in the wax, and let out all the fecess, but when the ſpirit quinteſſence comes, ſtop it ſuddenly with your finger, and turne it up.

The quinteſſence of all manner of Herbs, flowers, roots, fleſh, blood, or Eggs is thus done, take any of theſe things, and bray them in a Morter, or on a grinding ſtone, with fifteen times as much common ſalt, then put them being ſtopped in horſe dung (which muſt be reneWed once a week) for the ſpace of fortie dayes, by which time they will be putrified, and become water, then put it in a glaſs Still, and draw off as much as will come, then put that which came over and the feces together again, and braiſe them on aſtone; diſtill, and grind it often, then take the water and Circulate it in a Pellican, as before. To ſeperate all the foure Elements out of any of theſe things, put the putrified water into a Glaſs Still, and in Baln. M. draw of as much as will come Which is the element of water poure back that water, and ſet all together againe in Baln. M. to Circulate the ſpace of eight dayes, then ſet your Veſſell in aſhes or ſand fornace with a head, and receiver well luted, in a ſtronger fire then before, and a yellow water will riſe which is water, and ayre, put thoſe again in Baln. M. and

B b   the

*the water only will rise, so you have them seperated, then to
every pound of matter remaining in the bottom of the Glass
put four pound of Water, and set it as before to digest in Bal.
Ma. eight daies, then set them in a sand fornace, and give
a strong fire, and there will ascend a red Water Which is
fire, and Water, seperate them as before the water and ayre
so you have all the foure Elements asunder, Calcine the
earth, and rectifie the other Elements as you did for the
quinteffence.*

Aurum Potabile.

MAk an Amalgama of Sol and Mercury, vapour away the
Mercury, then take that Sol being in fine pouder, and heat
it upon a Plate of Iron till it be red hot, and quench it in
℞ but be sure to stop the Veffel suddenly, that too much of the
℞ burne not away do this fiftie times till the ℞ be tincted
yellow.

Another way.

TAke your Aurum in powder as before, cast the pouder into diftilled Vineger or Vrine, set it in the hot Sun, and you
fhall fee it rife in a thin Filme or Skin, take that off with a
feather, and put it in a Glafs where water is, do this as often as
any will rife, vapour away the water and the oile will remaine.

The quinteffence of Quick-filver.

TAke fublimate, and difolve it in *Aqua fortis*, diftill of the
water, and then the quinteffence of Quick-filver will follow,
rifing white as Snow, put back the Corafive water upon the
feces left behind, untill all be come over.

The quinteffence of Brimftone.

TAke pouder of Brimfton, and put it in old Urine, fet it upon a
foft Fire, till the Vrine be coloured, then poure of that, and
put

Natura Exenterata, page 376. By permission of The Huntington Library, San
Marino, California.

Chymicall Characters.

24 Grains   one penny waight.
20 Grains   a Scruple.
3 Scruples   a Dram.
60 Grains   a Dram.
8 Drams   an Ounce.

Saturn, Lead.   To purifie.

Venus, Copper.   Salt Peter.

Mercury, quickſilv.   Salt.

Antimony.   Salarmoniake.

Arcenick.   Tartar.

Allum.   Blood.

Vinegar.   Fyre.

Vinegar diſtilld.   Ayre.

Aqua fortis.   Water.

Aqua Regis.   Earth.

Wax.   Day.

Pot-aſhes.   Night.

Oyle.

Natura Exenterata, page 380, "Chymicall Characters." By permission of The Huntington Library, San Marino, California.

Lady Anne Clifford (author's own collection).

Margaret Cavendish, *The Philosophical and Physical Opinions, written by her Excellency, the Lady Marchionesse of Newcastle.* By permission of The Huntington Library, San Marino, California.

Here lyes the Loyall Duke of Newcastle and his Dutches his
second wife by whome he had noe Issue her name was Margaret
Lucas yongest sister to the Lord Lucas of Colchester a noble familie
for all the Brothers were Valiant and all the Sisters virtuous This
Dutches was a wise wittie & learned Lady which her many Bookes
do well testifie she was a most Virtuous & a Loveing & carefull wife & was
with her Lord all the time of his banishment & miseries & when he
came home never parted from him in his solitary retirements.

Margaret and William Cavendish. Copyright: Dean and Chapter of Westminster.

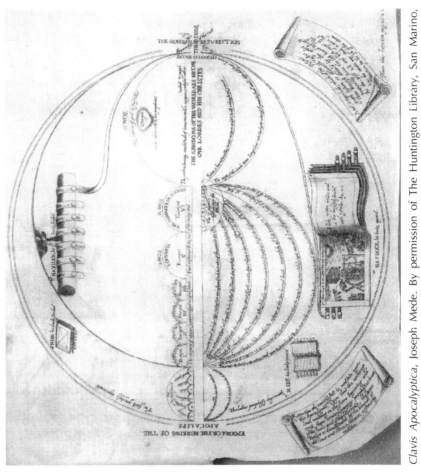

Clavis Apocalyptica, Joseph Mede. By permission of The Huntington Library, San Marino, California.

Chapter 5

Healers Extraordinaire: Seventeenth Century Alchemical Sisters and Their Recipe Books

Alchemy has no other kinship with chemistry than that their names are similar. True, alchemy is the mother of Chemistry, but can the daughter help it if her mother is mad?[1]

The following women represent what I like to think of as the healer-alchemists. Including these women in a book that addresses both physical and spiritual alchemy can be problematic. Some traditional historians of science and alchemy label their work iatrochemistry or refer to it as pseudo-chemistry. Iatrochemistry is a term that was coined by historians to refer to the healing aspects of alchemical practice but has in time become written about as if it is separate from alchemy. Alchemists such as Paracelsus, Basil Valentine, and Agricola were also associated with iatrochemistry.

Wolfgang Schneider suggests that the distinction that arose between alchemy and iatrochemistry may have its roots in a mistaken belief by some historians that iatrochemists did not experiment but merely disseminated medicines.[2] Marie Boas defines iatrochemists as those who did not seek the Philosopher's Stone. They might seek transmutation but in the spirit of making medicines.[3] However, that is not consistent with the alchemists, male or female, who did experiment with all types of materials associated with alchemy. This does a disservice to understanding the history of alchemy. Furthermore, as I have argued in the introduction, supported by the work of scholars such as Hutton and Hunter,[4] many women who studied alchemy did so in order to create healing ointments, medicines, powders, and tinctures that would ease the suffering of both their loved ones as well as their household staff. They read the works of well-known alchemists and included their recipes in their collection. They traded recipes with their peers and handed the books down to family members. Their work was sometimes published as recipe books, which most likely explains why they have been overlooked. It does not seem likely in traditional alchemical studies to cite evidence of the *Magnum Opus* in books that also give suggestions on the best way to prepare rabbit.

Although my focus in this book is on women, it is significant to note that they were not the sole writers of recipe books, either alchemically inclined or otherwise. Giambattista della Porta's 1658 *Natural Magik*[5] and Robert Boyle's *Medical Experiments: or a collection of choice and safe remedies for the most part simple and easily prepared: very useful in families and fitted for the service of country people*[6] are very similar to the women's recipe books with a few exceptions that I will point out later in this chapter. In fact, there is a distinct difference between the recipe books of the women alchemists and those written in a manner that we might more typically associate with the idea of a recipe book. The comparison will highlight why I agree that evidence for women's alchemy can sometimes be found in a non-traditional genre of writing.

Marie Meurdrac: *La Chymie*

Marie Meurdrac is described by historian, Lucia Tosi as the first woman to publish a book on alchemy or early chemistry, *La Chymie charitable et facile, en faveur des Dames Charitable* [*Easy Chemistry for Women*], 1656.[7] The majority of my comments regarding Marie Meurdrac have been obtained from the French edition of *La Chymie* edited by Jean Jacques in 1990[8] for which I have an unpublished translation.[9] Very little biographical information is known about Marie Meurdrac. Jean Jacques writes that after much research it appears that Marie was the older sister of Catherine Meurdrac who married to become Madam de la Guette (b.1613) who wrote in her memoirs about her older sister's work. Their father was a minor noble from Contentin who had married Elisabeth Dovet from Paris. Marie was an accomplished chemist.[10]

Publication of *La Chymie* was not easy for Marie. She did not have much self-confidence regarding her abilities as either a chemist or a teacher. Marie Meurdrac was a reluctant author, concerned that those who believed that women should not write or teach would criticize her. In her foreword, Marie states that she finally decided to publish her work in the service of healing in God's name. Yet, Marie Meurdrac also exhibits that spirit typical of the women alchemists as she reminds the reader that if the same amount of money and resources were dedicated to the education of women, they would succeed equally with men. In her words, "minds have no sex."[11]

La Chymie not only gives detailed instructions for medicines and cosmetic ointments, but Marie exhorts her readers to be sure to do as she does, distribute these remedies free of charge to the poor, a common practice shared by most of the women alchemists discussed in this book. She also offers to teach women in her own laboratory if they feel unsure about attempting the work. Lucia Tosi reminds us that in Marie Meurdrac's day, she labored under the constraint that a woman could not sell medicine and should keep a low profile in public. Furthermore, in 1551, an edict had been issued in France that forbade working with a furnace or with metals without the King's permission. Thus, the fact that Marie had both access to a furnace and was allowed to experiment as she did indicates that she must have had permission from the King.

The women I have been discussing in this book might seem as if they were the norm but we must remember that the education of women was still suspect in the seventeenth century. The French debate over women's education was called "*la querelle des femmes.*"[12] A quote from Marie's foreword illustrates rather poignantly her lack of confidence in her right to stand beside her male counterparts.

> When I began this small treatise, it was for my sole satisfaction, so as not to lose memory of the knowledge that I had acquired by means of long toil, and by divers experiments repeated several times. I cannot conceal that seeing it achieved beyond what I had dared to expect, I was tempted to publish it; but if I had reason to bring it to light, I had even more reason to keep it hidden and not to expose it to general censure. . . . I dwelt irresolute in this combat almost two years. I objected to myself that teaching was not the profession of a woman; that she ought to remain in silence, to listen and to learn, without bearing witness that she knows: that it is above her to give a work to the public, and that such a reputation is not by any means advantageous. . . since men always scorn and disapprove of the products that come from the mind of a woman. Moreover that they do not want the secrets to be revealed; and that finally it might be found, perhaps, in my manner of writing, many things with which to find fault. I prided myself that I am not the first woman to have placed something under the press, that mind has no sex, and if the minds of women were cultivated like those of men, and if we employed as much time and money in their instruction they could become their equal.[13]

Marie begins *La Chymie* with a dedication to the Countess of Guiche who appears to have had a connection to Louis XIV (most likely how Marie obtained permission to use a furnace). Marie explains to the Countess:

> Based on these foundations, I take the liberty, Madam, to present this small fruit of my vigils to you: it is in the interest of the conservation of your health, since it will give you quantity of remedies to contribute to it. If it has the honor of pleasing you, I assure you that it is true and faithful; and that my greatest passion always was to testify to you with its submission. I am, Madam, your very humble and very obedient servant, Marie Meurdrac.[14]

Marie's recipes follow a typical format for the recipe books of the day. *La Chymie* contains recipes for simples, tisanes, tinctures, and oils for headaches, ear pain, melancholy, kidney stones, and to guard against the plague. The recipes can assist with childbirth, dropsy, and provide needed purging. Complaints include jaundice, dysentery, diarrhea, fever, eczema, and various wounds. She also gives directions for making Macrobes' Pills that she states may help people live to 100 years of age. Our modern ears may find it amusing but she wrote it in all sincerity and it was likely taken that way by her readers.

They evacuate and consume by this means all peccant and superflous humors, strengthen natural warmth and all internal faculties; they postpone old age, since those that use them, ordinarily live to one hundred years; they maintain health and vigor, they relieve colds and headache, preserve vision, clean the stomach, protect from stones and caculus, and are also and great remedy and preventative in times of plague, and against epidemic illnesses.[15]

Marie's work details the operations that we associate with alchemy. She discusses the relationship between sulphur, mercury, and salt. She also references noted male alchemists such as Raymond Lull and Basil Valentine. Her work was in the medicinal tradition of Paracelsus and mainly focused on the use of vegetable materials. The belief was that since the vegetable kingdom was created prior to animals (according to Genesis) and that vegetable matter was not destroyed in the great flood, it was superior to all other matter.[16]

Marie's remedies that were based on the animal kingdom veered toward the fantastic. She subscribed to sympathetic healing where the germ of the curative is found in the cause of the disease. For example, she explains that if a viper bites someone, eating the snake will assist the cure. Crushing the insect that causes a sting and mashing it over the wound would heal it. Cutting a live chicken in half and placing it on one's head was a remedy that "fortifies the brain, and stops the eccentricities that come from a violent fever."[17]

Evidence of Marie Meurdrac's alchemical knowledge is especially apparent in her commentaries on the operations used in the preparation of her medicines. As well, she explains some basic information about the elements for the reader. Marie puts alchemy (or chemistry) of the day in plain words, explaining how the three principles of sulfur, salt, and mercury work together.

Marie's recipes that focus on the use of minerals and metals make clear that she was familiar with typical alchemical essentials such as Romain Vitriol (ferrous sulfate), nitre (potassium nitrate), saltpeter, sulfur, antimony, tartar and their various uses. She warns the reader that mineral and metallic curatives are very strong and she hesitated to include them in her book.

Although metals and minerals seem to be far removed from man, and since the Holy Scripture makes no mention of their creation, nevertheless they do not fail to provide us very healthy medicines. Today Medicine makes use of them with happy success. It is necessary that their preparations are made exactly, in as much as they are violent remedies; although one takes them only in minor amount, and for resistant and chronic diseases. When I started this book, I intended to not pass on my experiences. *That is why I omit in this part the operations about gold, and about silver, not knowing their preparations, or their use in medicine.* [Emphasis mine] I have seen several operations which are given the name drinkable gold, tincture of gold, oil of silver, that I could not understand; which could not persuade me that such perfect and condensed bodies, could be liquefiable. It is not that I condemn these operations because I cannot conceive of them; I would be as foolhardy as blind men, who would assure that there would not be Sun, because they would not see it. For the operations that follow, I assure that they are true and tested.[18]

Nevertheless, Marie includes the following discussion of the use of vitriol.

The virtues of the spirit of vitriol are great. It is notable that it should never be taken alone, and that its dose never exceeds three or four drops . . . It moderates the intense heat of malignant and violent fevers, and consumes the decay of the humors which cause them. It purifies blood, and penetrates into the veins: It is a diuretic; it kills the worms . . . raises the cankers, and cures the ulcers of the mouth. It is necessary to take guard that it touches no other part but the bad, since it corrodes the flesh: it bleaches the teeth if one rubs it on them with a small flag [cloth]: it helps to extract the tinctures from all kinds of flowers.[19]

La Chymie is an extraordinary collection of medicines, directions for healing, information on natural philosophy, and commentary on the effectiveness of different curatives. Marie Meurdrac provides the reader with all she will need to create her own apothecary. Needed implements and essential ingredients are described and explained. Finally, Marie exhorts her readers to be brave and confident practitioners of the ancient arts of healing. The following woman alchemist was an Italian. Like Marie Meurdrac, very little is known about her other than her work.

Isabella Cortese and Her Secrets

Who was Isabella Cortese? Her recipe book, *I Secreti*, brought her alchemical practice to the notice of historians.[20] *I Secreti* was first published in 1561 and then updated in 1565.[21] There is very little known about Isabella Cortese. However, the fact that she was able to publish her work suggests that she was certainly literate, probably relatively wealthy, and familiar with both medicinal and mineral alchemical work. William Eamon mentions that Isabella learned alchemy from many sources from her travels, learning "techniques firsthand from alchemists in Italy and Eastern Europe."[22] He adds that Isabella was skeptical of alchemical theory although she was serious about the art. It appears that her disagreement was more about procedure than in the belief of transmutation. "If you want to practice the art of alchemy, she warned readers, 'don't follow the teaching of Geber or Ramon [Lull] or Arnaldo [Villanova] or any of the other philosophers, because their books are full of lies and riddles.'"[23]

I Secreti opens with the first chapter addressing plague and poisoning. The first recipe concerns an oil "made by Brother Gregorio Mezzo for Pope Clement VII."[24] She states that the oil would protect "against poison and plague, and was tried on two prisoners in the Capitol, who were sentenced to death, and did the test." It appears that one of the prisoners died anyway but one survived, relating his ordeal to a friend of the Pope, Senator M. Simon Tornaboni.

Cures or prophylactics against the plague were a mainstay of alchemists who focused on medicine. Isabella continues with very detailed directions on the method. It involves taking the plant leaves and boiling them for four hours in the "bagno maria"[25] or Bain Marie which the reader will remember is a water bath. One continues with a number of steps that include the use of scorpions, thistle, and white dittany among others. The concoction is boiled, steeped in the sun, macerated, and cooled several times. She adds a section on the "Virtues of the Oil"[26] which Isabella assures the reader will counteract poison and if rubbed on the heart area and wrists, will protect from the plague. The oil would also protect if one was bitten by a rabid dog or injured with a poisonous weapon of some sort.

Isabella includes over 200 recipes that discuss not only working with plants but with metals. She writes of melting white iron and tempering. On another page, Isabella includes illustrations of typical alchemical equipment. She writes, "Questi sono i uasi della detta opera"[27] which translates very roughly to "These are used for the work." The illustration shows typical alchemical apparatus such as: "Forno, Vaso di terra, Recipiente, Ritorta, and Capello"[28] which again roughly translate to oven, vase/jar of earth, container, retort, and hat (looks like a cover for another apparatus). Isabella provides another drawing on how to heat "salnitro, tartaro, e zolfo che fissa l'amalgama."[29] The small retort fits into a larger one, both laying on their side with the smaller retort placed over the fire. Saltpeter or niter as well as tartar and sulphur (zolfo) were often used in alchemical work and it appears that she is explaining how to combine the elements in a fixed state.

I Secreti is a remarkable collection of all sorts of useful remedies that regardless of country of origin fulfill similar hopes for healing. The following recipe book is quite similar and extraordinary.

Alathea Talbot, Countess of Arundel: *Natura Exenterata*

My current students, candidates for a California teaching credential, must be tested for exposure to tuberculosis before working in a classroom. Actually all teachers and school staff make a visit to their doctor every three years to ensure school children are not exposed to a disease that was such a scourge prior to the 1950s and often known as *consumption*. Like many of the women alchemists, Alathea Talbot sought remedies for disease, both mild and severe. Her husband, Thomas, was stricken with consumption and they traveled Europe searching for a cure. However, first we need to get a better sense of who Alathea was as a woman.

Alathea Howard neé Talbot, future Countess of Arundel and Surry was born in 1588. Her mother was Mary Cavendish (daughter of Elizabeth Barlow or Bess of Hardwick Hall and Sir William Cavendish). Alathea's sisters, Mary and Elizabeth married respectively, William Herbert, Earl of Pembroke and Henry

Grey de Ruthin, who eventually inherited the title of Earl of Kent. The sisters were close friends of Queen Henrietta Maria and in fact, in 1642, Elizabeth was the one sister who joined Henrietta Maria in exile in the Netherlands.[30]

The Talbot family was embroiled in the politics of the day. An earl was beheaded, their title to Norfolk was seized, the son-in-law, William Herbert, managed to get one of Queen Elizabeth's ladies-in-waiting pregnant, refused to marry her, and thus spent some time in the Tower as punishment. Upon marrying Thomas Howard in 1606, Alathea Talbot became the Countess of Arundel and Surry. Thomas Howard's family was also no stranger to political problems with Thomas' grandfather (also named Thomas) having been executed for treason in 1572. Thomas' father, Philip, was also suspected of treason, probably due to the anti-Catholicism prevalent during Elizabeth's rule. He was imprisoned in the Tower of London, convicted of treason, and remained in the Tower until he died in 1595.[31] Therefore, Thomas grew up with both his grandfather and father convicted of treason; one was executed whereas the other died in prison. The family's property and titles were seized and needless to say, they had little money and were bitter. "From earliest childhood he [Thomas] was taught that the titles and properties that were his birthright, as heir to the greatest noble family in England, had been taken from him through two corrupt judicial sentences."[32] Thomas's title and property were restored in 1604 and he remained on good terms with the royal family. Thomas assembled a vast collection of art, 3000 books, medallions, and carved gemstones. His collection included pieces by Leonardo da Vinci, Michelangelo, Raphael, Van Dyke, and Reubens.[33]

Thomas was not immune to political ambition and was sworn into the Privy Council in 1616. In the interest of moving up the ranks of those in power, Thomas received communion by the Church of England despite his Catholic background. He rose to great power that waned when Charles I became king. By 1632, Thomas was found to be back on good terms with Charles. He is buried at Arundel Castle but his heart and viscera were buried in Padua at St. Anthony's basilica.

By 1612, Thomas had contracted consumption and went searching for a cure. He traveled in continental Europe for a short time, returned home, and then set off with Alathea to Europe, again to find a cure. Eventually, the couple spent time at the hot springs of Albano, near Padua. Thomas and Alathea actually lived in a monastery for a time while focusing their intellectual energy on increasing their fluency in Italian. They returned to Britain in 1615 upon Thomas inheriting property from an uncle. Furthermore, Alathea's father died in 1616 leaving her and the family with ample funds. Thomas and Alathea had a falling out (sometime in the late 1630s) due to her pro-Catholic stance. He lived by himself on the continent from about 1642 until his death in 1646.

Natura Exenterata

Alathea's notable book of recipes for all manner of needs, *Natura Exenterata*, was published in 1655, apparently posthumously as her death is listed as May 24, 1654. She states on the title page that she has collected 1,720 receipts! She also includes a "A Catalogue of such Persons of Quality, viz. Knights, Doctors of Physick, Gentlemen, Countesses, Ladies and Gentlewomen, & c. by whose Experience, these Receipts following have been approved."[34] Some of the names of her contributors are familiar for the time such as Kenelm Digby, John Digby, Paracelsus, Mrs. Jones (very likely Katherine Boyle Jones, Lady Ranelagh), Mrs. Conway (again most likely Anne Conway) and Sir Walter Raleigh.

The recipes focus on instructions for preparing elixirs and powders for curing typical maladies of the day such as dealing with worms, fever, lung ailments, complications of childbirth, cancer, and a host of other diseases. Alathea's medical chemistry follows the same procedures as the alchemical processes practiced by her male counterparts, although her focus is practical healing and not overtly spiritual. A quote from her chapter on "Chymicall Extractions"[35] sounds precisely like a reading from Paracelsus or van Helmont. The only change I have made is to write the middle English "f" as "s" to facilitate reading.

> Take [symbol I cannot identify] seven times rectified, and put it in a pelican, and let it Circulate forty dayes and this is his burning water or quintessence or Heaven, and hath a smell so odoriferous that it being open in any Roome, all that passe by will be forced to come in, if the smell be wanting when you open it, put it in a bolts head well stopped with wax, and bury it in Horse dung with the bottom upwards, some part being bare to the Ayre, and having stood so many dayes, take it out, and with a small warm wyer, bore a hole in the wax, and let out all the feces, but when the spirit quintessence comes, stop it suddenly with your finger, and turne it up.[36]

Alathea declares that this recipe is the best method for extracting the quintessence from herbs, flowers, blood, and eggs. The quintessence described in alchemical literature typically refers to the most basic and pure element of the ingredient being used. When the impurities are distilled and burned away, the essential substances are left behind. Martinus Rulandus defines the *quinta essentia vegetabilium* as "that which is extracted from the components of vegetable things."[37]

Any reader of alchemical literature will recognize Alathea's procedures. She references use of the *Balneum Marie* (from Maria the Jewess), the retort, pelican, crucible, and athanor, as well as the operations of dissolving, digestion, distillation, separation, and calcination. She heats her matter in the fire as well as in horse dung, a common alchemical practice. Distillation might make use of vinegar as well as urine. She mentions that in a successful process:

There will ascend a red water which is fire, and water, separate them as before the water and ayre so you have all the foure Elements asunder, Calcine the earth, and rectifie the other Elements as you did for the quintessence.[38]

Alathea gives directions for making Aurum Potabile (an alchemical elixir made with gold), where she instructs the reader to "Make an Amalgama of Sol and Mercury."[39] Her language is the language of the *Magnum Opus*. She discusses extracting the quintessence of quicksilver as well as working with antimony, indicating that she was no stranger to working with metals. The operations took several days to complete. She has cures for just about anything from bruises, cuts, and aches to coughs, worms, and broken bones. She also has advice for breeding horses and making dyes and laces.

Natura Exenterata also provides the reader with two different tables that explain the meanings of symbols used in the work. The symbols are typical of alchemical literature. For example, earth, air, fire, and water are the usual variations on a triangle. The direction of the point indicates the propensity of the element to move away from the earth or toward it—up / down, heavy / light.

| Earth | Air | Water | Fire |

Figure 5.1: Traditional Alchemical Symbols for Earth, Air, Fire, Water as Depicted in *Natura Exenterata*.[40]

A few of the other characters found in *Natura Exenterata* are given in Figure 5.2.

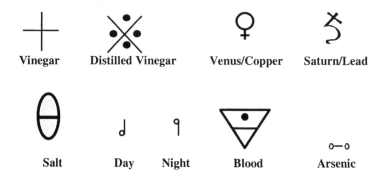

Figure 5.2: Traditional Alchemical Symbols for Common Elements as Depicted in *Natura Exenterata*.[41]

Alathea Talbot was also a contributor to another book of recipes and chemical work titled *The Queen's Closet*,[42] published for Queen Henrietta Maria in 1683. I believe Katherine Boyle may also be a contributor, identified by the author as Mistress Jones. (We know that Katherine Boyle developed many cures that were published by various authors). This collection of remedies was written in three sections:

> I. the Queens Cabinet Opened or the Pearl of Practice. Acurate, Physical and Chyrugical Receipts, II. A Queens Delight Or, the Art of Preserving, Conserving, and Candying. As Also, a Right Knowledge of Making Perfumes, and Distilling the Most Excellent Waters, and III. The Compleat Cook: Expertly Prescribing the Most Ready Ways, Whether {Italian, Spanish, or French} for Dressing of Flesh and Fish, Ordering of Sauces or Making of Pastry.[43]

Alathea Talbot's book provides more overt evidence of her alchemical knowledge than most, something we will also find with the next alchemist.

Lady Katherine Jones née Boyle, Viscountess Ranelagh: A Force of Her Own

Katherine Boyle Jones or Lady Ranelagh actually falls into two categories for finding evidence of women's alchemy. Her story is an exceptional example of the underlying role played by alchemy and its nexus with Apocalyptic study and the practice of Jewish mysticism or Kabbalah. However, in this chapter I will focus on the healing she practiced and wrote about in her recipe books.

The seventeenth century experienced a paradigm shift in which a movement from the medieval notions of reality to the modern sense of a scientific world-view emerged. The period of the Scientific Revolution is described as a watershed era.

> This was the century in which the English scientific community developed and took its place at the forefront of European scientific investigation. Indeed the English scientific community supplied a number of crucial elements that would unite to create modern science: a significant number of innovative and insightful thinkers, a national stake in a new methodology that stressed cooperative experiment and objectivity, and a gentlemanly coterie of scientists who applied their social standards of behavior to the ideology of modern science.[44]

Although social mores were shifting alongside the Scientific Revolution, women's roles continued to be marginalized. Nonetheless, some women toiled away at studying the natural world. For example, Anne Finch, Viscountess of Conway (1631-1679), developed her "theory of monads"[45] that her correspondent, Wilhelm Leibniz, used in his own theory of the universe. Margaret Cavendish, Duchess of Newcastle (1617-1673) wrote on natural philosophy and was begrudgingly invited to attend a meeting of the Royal Society. Their stories will be discussed in a subsequent chapter. Katherine Boyle appears to have successfully balanced her scientific interests with the expectations for how a woman should act in her society; at least, she did not seem to garner the criticism of her peers as many other women alchemists did.

Katherine Boyle was born on March 22, 1615 in Ireland. She was the fifth daughter of Richard, the very wealthy 1st Earl of Cork (1566-1643). Richard is said to have begun his political ascent under somewhat dubious circumstances in Dublin and his character left something to be desired although he was quite clever.

> His numerous sons and daughters copied not so much the man Cork really was, but the myth he created about himself . . . his shrewd intelligence, consuming ambition, and obstinate perseverance in any task he set for himself. . . . Cork pretended to possess—honesty, piety, and charity.[46]

Richard's first wife, Joan Apsley, died giving birth to a stillborn son in 1599. Richard's second wife, Catherine Fenton (c. 1588-1630), was a woman he eventually loved deeply. Catherine Fenton's father was Sir Geoffrey Fenton, principal secretary of state for Ireland. Richard and Catherine had 14 children with Katherine being seventh in birth order.[47] Richard believed that boys should be schooled carefully, whereas girls were to be married off to increase his family holdings.[48] When Katherine was only six years old (1621), Richard arranged a marriage for her to a boy, Sapcott Beaumont, whose family resided in England. Upon reaching the age of about eight years (1623), Katherine was sent to England to live with her intended's family. Katherine returned to Ireland in 1628 at the age of 13 years. The marriage never took place; however, Sarah Hutton suggests that perhaps Katherine's five year sojourn in England was where she received a rather comprehensive education. Katherine's source of education is significant because as I mentioned, her father did not believe that women should be educated in subjects other than stitchery, dancing, and skills needed to be a useful wife according to upper class mores. Katherine, however, downplayed her education for some reason; yet, she was described as one of the most intelligent women of her time, earning a significant reputation for her intellect and mental acuity. Sir John Leake, a friend of Katherine's sister, Alice, described Katherine as "'a more brave wench or braver spirit you have not often met whithal. She hath a memory that will hear a sermon and go home and pen it after dinner verbatim.'"[49]

Katherine's father arranged all of his daughters' marriages with money and status in mind. He ended up choosing husbands who were less than congenial. Katherine's husband, Arthur Jones, was known to be a "drunken lout."[50] Her sister, Lettice, ended up with a husband who was cruel to her whereas another sister, Alice, married a man who showed her unending disrespect. The husband of their sister, Dorothy, was known to be unkind and Joan's husband deserted her for a while. Mary was alone in marrying someone with whom she fell in love, at least at first. Unfortunately for the Boyle sisters, Richard made poor choices for their husbands.

When his wife, Catherine, died in 1629, a heartbroken Richard moved the whole family to Stalbridge (in Dorset) and never remarried. Robert Boyle noted in his autobiographical notes that his father was inconsolable and that after Catherine's death, he marked the day each year. Catherine must have been an effective mediator between the tempestuous Richard and his children. Robert also described her loss in his notes as a "disaster."[51]

Katherine married the above-mentioned Arthur Jones, Earl of Ranelagh in 1630.[52] The couple began their marriage residing in Ireland but moved to London in about 1631. Their first daughter, Catherine, was born in 1633. Catherine was followed by Elizabeth who was destined to create a family scandal in 1677 by marrying a footman. Two sons, Francis and Richard (the future 2nd Earl of Ranelagh), were born in 1639 and 1641 respectively. Katherine's marriage is described as troubled.

> The Jones match does not appear to have been a happy one. Her husband had a reputation for boorishness, even at the time of their marriage. Katherine's residence in London during the 1640s and 1650s was probably, therefore, a separation of convenience, as well as a refuge from troubles in Ireland.[53]

Unfortunately, Richard followed in his father's footsteps. Although he was given the post of Paymaster-General, Richard was convicted in 1703 of stealing an astounding £72,000!

Katherine was close to her siblings. Fortunately, one sister, Mary Boyle Rich, who became the Countess of Warwick upon her marriage, kept a wonderful little diary that sheds light on their relationship. Fell Smith introduces the work observing, "Nearly ten years older than Mary, she [Katherine] remained all her life the closest and dearest friend of this youngest sister. In Mary's Diary she is always 'Sister" or 'Sister Ranelagh,' sometimes only 'S. R.'"[54]

Mary Boyle Rich

I wish to include a brief account of Mary at this point in the chapter since her story is directly tied to Katherine's alchemical and religious beliefs. Mary had been the sole sister to defy her father's choice for marriage. After soundly rejecting James Hamilton, heir of Viscount Clandeboye, a difficult power struggle erupted between Mary and her father. He tried withholding her allowance and invoking the support of his other children but to no avail. Mary would not give in. Richard arranged another marriage with Henry Moor, son of Lord Moor with whom he hoped to solidify an alliance. This marriage, too, did not take place but it is unclear as to who broke off the engagement. Mary prevailed and was given her allowance as well as that owed her previously.[55]

Mary met Charles Rich in 1639. He was 23-years-old and handsome; however, being the second son of the Earl of Warwick meant that it was not a good financial match. Nonetheless, Mary's sister-in-law, Elizabeth, assisted Charles in his romancing of Mary. She was not a pushover but Mary finally consented to the marriage. The convincing of her father was no easy matter and amidst much drama, Mary finally declared that she would not marry Charles without her father's consent but she would marry no other man and remain a spinster. Richard Boyle had met his match and gave in. Refusing to participate in a large, public wedding, the couple married privately, a ceremony that occurred without the knowledge of their respective parents. The entire Boyle clan except for Katherine shunned Mary.

Due to their limited finances, Mary and Charles lived with his parents until his father's death. Their first child, Elizabeth (1642), died as an infant at 15 months of age. A son, Charles, was born in1643. Interestingly, it appears that the couple decided to have only the two children due to their limited income and Mary's wish to avoid the effect of continued pregnancies on her figure. However, their method of birth control is not clear.

Life with the Warwick parents was ruled by Puritan practice although Mary was ambiguous about religion at the time. Yet, she began to be influenced by the constant exposure to Puritanism as well as being exhorted by her siblings, Katherine and Robert, who were in the Millenarian movement by now. They must have feared that since they believed that the Second Coming of Christ was imminent, their feisty younger sister was in danger of being remanded to hell.

Charles became quite difficult to live with after he contracted gout. When their son, Charles, nearly died in 1647, Mary believed it was punishment for her sins. She made a bargain with God that if her son would be allowed to live, she would repent. Her son did get better, at least for a time; however, he died of smallpox in 1664 at the age of almost 21 years. An entry in Mary's diary on the anniversary of her son's death reads:

> May the 16[1667] I kepte it a private fast being the day three yeare upon which my son [Charles] dide . . . had . . . large meditationes upon the siknes and death of my onely childe . . . his sik bed expressiones . . . how god was pleased to waken him . . . then I begane to consider what sines I had comited that should cause God to call them to remembrance and slay my sonn.[56]

Mary believed that she had met her match with the Divine and from that time on, she immersed herself entirely in a spiritual life of prayer and study. Mary's diary begun on July 25, 1666 records her tumble into obsessive piety, blaming all misfortune on her earlier behaviors as a young woman who rather than seeming spirited, was now characterized in her mind as having been willful, sinful, and vain.

The British political scene was in turmoil during these years with the civil wars taking place, culminating in the execution of Charles I. The Warwick family walked carefully between their loyalty to the Crown and to parliament. Upon the death of both his older brother and uncle, Charles Rich became Earl of Warwick in 1658/59. This changed their financial situation significantly. Mary and Charles tried to have more children but they were by now 39 and 48-years old respectively. Mary believed her inability to conceive was punishment from God for having practiced birth control as a younger woman.

Mary was unusually self-critical, constantly holding herself up to an impossible ideal that she could only fail to achieve. In her work titled, *Occasional Meditations*, Mary writes of the beauty of nature as well as about her pet dog, birds, and ethics. However, in her diary, she turns a supremely critical eye on herself. The entry for Christmas 1676 is a poignant example.

> Those sins which, in an especial manner I bewailed, were my heart sins, and my breach of my baptismal vows, and my sins against gospel light, and my unprofitableness under the excellent means of grace I had so long enjoyed, and my want of life and vigour in holy duties, and the great vanity and inconstancy of my mind, and my unequal walking with G. [God], and my backslidings from Him, and my breach of my promises, and my too much love of the world, and my too little prizing of Christ and my want of zeal for His glory, and the great vanity of my thoughts and words, and my great unusefulness in the place G. hath set me in, and my not improving the opportunity of doing and receiving good, and my great unthankfulness, and my crimson sins against mercies.[57]

Mary's piety was echoed by both Katherine and her brother, Robert, who at an early age evidenced a certain dread regarding his own salvation. Mary loved Robert deeply and enjoyed his visits to her home as well as those from her sister, Katherine. Although the focus of this chapter is healing recipes, it is important for the reader to understand that theology was often embedded in that same work. Healing would occur with the grace of God and did not suffer the line that separates the secular from the non-secular world that we see in modern times. This becomes even more evident as we explore Katherine's studies in the following sections.

Sister and Brother: Best Friends

Robert and Katherine were actually best friends as well as siblings. Robert (b. January 25, 1626) had traveled as a young man in Europe. Robert had some peculiar beliefs that may have emerged out of his religious views. He stuttered and believed that it was a result of his making fun of some children he knew who stuttered. He thought perhaps he was being punished for his callous words.[58] Robert and his brother, Francis, were sent from Ireland to be schooled at Eton in 1635. Richard Boyle's diary indicates that the Provost, Sir Henry Wotton (1568-1639) was a close friend of his. As Robert describes his education and tutors, it is very clear that his father believed educating his boys was very important and it seems that Robert was his favorite.

Upon returning to England in 1644, Katherine persuaded Robert to maintain a life in the country rather than join the Royalist army. Robert moved to the family manor in Stalbridge, Dorset where he first tried his hand at writing but by 1649, he had built a laboratory at Stalbridge and by 1650 had refocused his energy on the study of the natural world. Robert moved to Oxford in 1655 and finally to London in Pall Mall with Katherine in 1668. 1670 was a difficult year for Robert as he suffered a severe stroke that almost killed him (he would have been only 43-years-old). Robert may have never married but it seems that he had positive relationships with the women in his life. He loved his nieces as much as they loved him.[59] Upon his death, Robert left much of his property to his nieces, ignoring the nephews, who could probably take care of themselves. His father, Richard, had done the same thing when he died which was consistent with his belief that women were fragile.

Katherine encouraged Robert's work which encompassed alchemy, physics, and theology. Their lives were intertwined both scientifically and socially. For example, their good friend, Robert Hooke, who is well known for his work in microscopy as well as numerous other scientific pursuits, was a frequent visitor to the Boyle household. Katherine was responsible for Robert Hooke building a lab for Robert at the Pall Mall house in 1676. Hooke's diaries provide a glimpse into his close relationship with the brother and sister. The years 1672-1680 show Hooke having dinner with Lady Ranelagh and Robert often. The entry for Friday, August 24, 1677 refers to the work on the Pall Mall laboratory. "To Mr. Boyles, dind with him, beggd long screw quadrant, and took mandrill. Directed Laboratory."[60] Hooke is known for his temperamental relationship with Isaac Newton and it appears in a few entries that he and Katherine were no strangers to arguing. Hooke recorded on Monday December 17, 1677: "I was at Mr. Boyles, Lady Ranalaugh huffed. I fitted Boyles microscopes for Insects."[61] Hooke notes a coolness when he sees the Boyles that Friday, December 21.[62] The next day, Saturday, he observes there is some reserve in their interactions: "With Boyle and Lady Ranalaugh againe about Louden. More puts off."[63] Apparently this dust up was still going on the next day but it is amusing that Hooke's diary shows that he still had dinner with the Boyles on a regular basis and their association must have been important in light of the number of references Hooke makes describing the state of the relationship. On Thursday, June 20, 1678: "not well. to Sir Ch. [Charles] Wrens, examind Kempsters bill. At Lady Ranalaughs, she scolded &c. I will never goe neer her againe nor Boyle."[64] Katherine must have been able to smooth things over; they continued to dine often despite Hooke recording every so often that Katherine was "still finding fault."[65]

Learning Hebrew was often accomplished by alchemists who wanted to read early writings and as well, by Millenarians who wanted to study writings on the Apocalypse. Katherine's Millenarian connection will be addressed in a subsequent chapter. *The First Gate, or, The Outward Door to the Holy Tongue, opened in English* was published in 1654. *The First Gate* is a Hebrew grammar book written to assist a student to decipher the language. The author, William Robertson encourages the reader stating in the title page that it is "all in so plain and easie way, as may be made use by any ordinary Capacity of either Sexe."[66] The book is dedicated to Katherine Boyle/Lady Ranelagh. Robertson was teaching her Hebrew and she was extremely proficient. Robertson acknowledges Katherine's hard work but also explains that he thought putting his teaching method into writing might be of use to others. He states that teaching Lady Ranelagh clarified and supported his theories that: a) women could learn Hebrew, b) one did not need to have Latin in order to learn Hebrew as they are so different, and c) that his "helps"[67] would be effective for the learner. Robertson also wrote *The Second Gate, or, The Inner Door to the Holy Tongue,* published in 1655.[68] Lady Ranelagh is again mentioned in the inscription where he explains that *The First Gate* was so well received, he believed students of Hebrew would find a dictionary (of sorts) useful. He provided translations of Biblical Scripture in Hebrew and English.

Katherine died on December 23, 1691; Robert followed a week later and they were buried side-by-side in the south chancel, St. Martin-in-the-Fields, London. A story was told that when Katherine died, people reported seeing a flame shooting out the chimney that could not be explained. When Robert followed his sister, the neighbors claimed the chimney flamed again—but it is only a story.

Katherine was a supporter of educational reform as was Dorothy Moore, her childhood friend who had become Katherine's aunt-by-marriage. Dorothy supported the notion of women becoming ministers; thus, exposing Katherine to ideas that were less traditional for women. Moreover, the Boyles seemed to have been able to navigate between Royalist loyalties and ties to Parliament.[69] Yet, despite Katherine's unhappy marriage, she must have valued the institution to some extent as Matthew and Harrison note that it was Katherine who convinced her best friend, Dorothy Moore, to marry John Dury.[70] Dorothy Moore will be discussed in chapter seven.

Katherine's Alchemical Work

Katherine Boyle's alchemical work is found in her recipe books, *Kitchin-Physick, WMS 1340* and in *My Lady Ranelaghs Choice Receipts, MS Sloane 1367.*[71] Katherine's connection to alchemy is quite logical given her living circumstances. It is known she worked with medicinal herbs and her brother practiced alchemy, both to attempt transmutation and to create medicines. Reading correspondence between Katherine and Robert demonstrates the ease with which they conversed on scientific matters. Their topics of discourse included politics, theology, and philosophy. Robert's letters often refer to Katherine's previous letters to him in which she has raised these issues indicating that the siblings entertained a lively exchange of ideas. For example, in an early letter dated March 6, 1647, Robert writes about one of his furnaces falling apart upon being moved which was a disaster for his experiments. He states:

> Well, I see I am not designed to the finding out the philosopher's stone, I have been so unlucky in my first attempts in chemistry. My limbecks [alembics], recipients, and other glasses have escaped indeed the misfortune of their incendiary, but are now, through the miscarriage of that grand implement of Vulcan, as useless to me, as good parts to salvation without the fire of zeal. Seriously, madam, after all the pains I have taken, and the precautions I have used, to prevent this furnace the disaster of its predecessors, to have it transported a thousand miles by land, that I may after all this receive it broken, is a defeat, that nothing could recompence but that rare lesson it teaches me, how brittle that happiness is, that we build upon earth.[72]

Carl Zimmer maintains that Katherine introduced Robert to alchemy, which is logical considering that she was associated with alchemists such as those in the Hartlib Circle, well before Robert settled in England and began his scientific pursuits. [73] Certainly, Katherine would have had been familiar with the tools of both alchemy and herbal work. This intertwining of alchemy/chemistry and medicine is repeated often in so-called, *women's alchemy.*

Lynette Hunter discusses the common occurrence of women carrying out kitchen chemistry and suggests that in the seventeenth century, as the study of science became more institutionalized, it became the province of men. Upper class men living in a patriarchal society could not have their work associated with the kitchen and moved it into the laboratory. Often, Katherine Boyle Jones and other contemporaries' medical receipts became included in books attributed to men such as Thomas Willis, Samuel Hartlib, Robert Boyle, Kenelm Digby, and John Evelyn (all Royal Society fellows) as opposed to being published under the women's own names, not an unusual practice at the time. Hunter writes:

> As medicine and chemistry become an activity for aristocratic men during the 1650s, those men needed a way of differentiating their work from that of their female counterparts, partly to avoid being trivialized, partly to enter the public realm, and partly because of a growing differentiation between gendered activities.[74]

The societal rules of the new science were also developed by these gentlemen scholars.

> Since he was a gentleman and since the codes of conduct concerning the new experimental science had not yet been established, Boyle and his gentleman colleagues took over the codes of ethics from their social position. A scientist's word became his bond, scientists collectively determined who could achieve the status of a scientist, and the polite discourse of the town club became the accepted mode for scientific conversation. More than any scientific discovery,

this development of rhetoric and behavior helped to create the science we recognize today.[75]

If one needs further evidence that Katherine approached her work from an alchemical paradigm, one only needs to examine *My Lady Ranelaghs Choice Receipts, MS Sloane 1367.* It contains the typical types of recipes I have been discussing such as, "My Lady Barringtons rare Bal(s)um for the Palsy, Headake, Paines in the Joynts."[76] However, Katherine's alchemical knowledge is more explicit.   She uses abbreviations for elements and alchemical operations throughout her work which she explains in a table at the end of the book. Her symbols follow typical alchemical notation for the time. The symbols that denote acetum (vinegar), aqua fortis (nitric acid), aqua Regis (mixture of hydrochloric acid and nitric acid), and tartar are just a few examples found in her table that are identical to those used by well-known alchemists such as Basil Valentine. Both Katherine and Alathea Talbot use the standard symbols for the elements such as copper, quicksilver, and lead as well as those used for earth, air, fire, and water. Furthermore, Katherine used a notation that I have not seen previously for the following directions.[77]

Put it over the fire   △   *Take it off the fire*   △°

Figure 5.3: Symbols for Directions from *My Lady Ranelaghs Choice Receipts, MS Sloane 1367.*

The triangle is the common alchemical symbol for fire. The small circle is elegant in its simplicity as it quickly instructs the reader about what to do with the mixture. Katherine's directions for the operations of sublimation, precipitation, distillation, and calcination clearly indicate that she was well versed in chemistry. She also made use of the *Maria Balneum*, retort, crucible, and alembic.[78] A recipe from *Kitchin-Physick, #40* further illustrates Katherine's use of alchemical operations in her work. I have used modern spelling to make the reading easier, also highlighting the alchemical terms in italics.

> Just a pound of Bay salt in as much water . . . then if you please you may strain it, and in a *head and body* or in a *retort* draw it off first with a gentle fire till [*distillation*] . . .increase the fire by degrees and give for the two or three last hours a considerable one that the *caput mortum* [*dead head*] may remain dry, after the same manner you may make salt . . .[79]

Directions for making *Balsam of Sulphur* are found in both recipe books, number 48 in *My Ladies Remedies* and number 254 in *Kitchin-Physick*. This is identified as a "remedy for consumption or any cough."[80] In *Kitchin-Physick*, it is followed by directions for making Spirit of Sal Armoniack and Ens Veneris. Sal Armoniac or sal ammoniac is described by Martinus Rulandus in the *Lexi-*

con of Alchemy (1612) as being useful for the complexion.[81] Ens veneris was used for the spleen.[82]

A Small Mystery

Katherine has drawn some puzzling symbols in the corner of about 90 recipes of the recipe book from the Wellcome Library, *Kitchin-Physick, WMS 1340*. A few symbols seem to cover page numbers although whether that is an artifact or intentional is unclear. Some of the symbols resemble typical alchemical notation, although they do not match any other alchemist exactly. I have researched the origin and meaning of the symbols as they must mean something to the author that is unclear to me. There is no obvious connection between the recipe and the symbol written in the corner. Since Katherine's studies were so widespread, I have examined various alphabets, symbol systems, and even Mason's marks that they would carve into their stonework—like an artist's signature. The symbols do not match Agrippa's alphabet, nor are they Hebrew, Latin, Greek, or any other ancient language that seems an obvious possibility.

The symbol on recipe number 31 contains two figures identified in *The Woman's Dictionary of Symbols and Sacred Objects* by Barbara Walker as the Cross Potent and a circular symbol for earth.[83] (see Figures 5.4 and 5.4a, and 5.5)

Figure 5.4: Cross Potent   **Figure 5.4a: Earth**

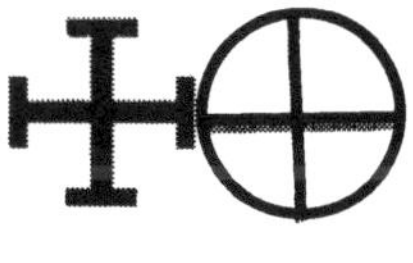

Figure 5.5: Combined Symbols As They Appear in *Kitchin-Physick*

Figure 5.4 can represent a crucible and in some traditions, Figure 5.4a is used to represent the sun and sometimes the Consecration Cross.[84] Combining symbols often denoted some compound; however, what Katherine meant by Figure 5.5 is unclear. There is another variation of this symbol used by the Millenarians, a group to which Katherine belonged and it will be discussed in chapter seven.[85]

Page number 11 of *Kitchin-Physick* shows a symbol that is associated with infinity (see figure 5.6). It is also associated with Kabbalah symbolism drawn with the rounded edges.

Figure 5.6: The Cross of Infinity[86]

Another unusual series of characters s uses variations of a symbol that was sometimes used by alchemists to denote water. (See figure 5.7) They span recipes number 73-81 and often have a number inscribed within which is what may have led some historians to conclude the purpose was to cover previous page numbers. Number 73 seems to have a number 3 inscribed within it. However, the character in number 77 looks like a number 3 turned 180 degrees which was often used to represent fire in alchemy.[87] The figure in recipe number 81 includes what could either be the number 8 or the symbol for infinity.

Figure 5.7: Early Symbol for Water[88]

Additional symbols in *Kitchin-Physick* resemble sigils used in occult studies and magic.[89] For example, the symbol for recipe number 17 resembles a similar sigil used to represent the moon. It looks like a cross or plus sign, +, with a small circle drawn at each arm.

Figure 5.8: Magical Symbol for Moon[90]

Many symbols resemble grimoires which might be drawn to assist the healer and the patient in the curative process.[91] They resemble faces of animals and even a few of the occult symbols for angels. If they were arranged in a particular way, the healer could call upon spirits for assistance.

Figure 5.9: Vertical Infinity

Interestingly, a symbol can be found in the corner of one of Robert Boyle's recipes found in the Boyle Papers, Volume 18, Folio 83v-84r, *Recipe mixing `the blood of a hermaphrodite and the blood of a sound Ruddy Comlexioned woman.*[92] The symbol on Robert's recipe does not match exactly but is very close to Katherine's symbols from recipes 49, 36, 39, 7, and 20b, which are variations of the vertical infinity symbol.[93] That symbol is identified as meaning a dram or an ounce in *Natura Exenterata*. Why it should be written in the margin would seem to suggest it is a direction for the reader or perhaps, the writer's own notation.

At first, I could not resolve Katherine's devout Christianity with the use of magical emblems. However, her brother Robert, whom we know was a very careful scientist, also studied the occult sciences "particularly witchcraft, spirit invocation, second sight, and magic" [94] The modern reader hears the term occult and thinks of superstition, magic, and witchcraft with the 21st century world view. Earlier in recorded history, the occult was believed as William Kingsland explains, to be a way to understand the workings of God's creation using one's inner reasoning and even openness to revelation. Kingsland contrasts modern science and occult science. Modern science came to focus on what is considered to be real and can thus be measured and/or manipulated. The occult sciences employed inner detection of the ways of the universe; thus, the individual needed to work on developing inner powers of perception in order to study and find answers.[95] Robert Boyle believed that the Philosopher's Stone could be used to call forth and communicate with angels. It appears that he wanted to prove the existence of supernatural phenomena as a means for combating the rise in atheism that he perceived in his day.

Mitchell Fisher also discusses the natural philosophers who were attempting to reconcile their scientific observations with theology. He notes that Copernicus, Kepler, and Gassendi still believed that God was the ultimate Creator and for example, the One who placed the sun at the center of the universe/solar system.[96] He reminds us how the 17th century thinker was trying to balance the new experimentation and theology. However, that small voice of new doubt was bubbling up in light of questioning and wondering how a miracle could actually happen scientifically. Fisher points out the dichotomy of the time when superstition was a partner to brilliant deduction. Using witchcraft as an example, Tudor laws were more lenient regarding witchcraft; however, James I who appears to have been phobic on the topic, resurrected the witchcraft laws and 70,000 men and women were executed as witches between 1600-1680. Whereas, Charles II employed a lab for experimentation and was well versed in alchemy. The same time period sees Newton and Leibniz developing the calculus.

Boyle is a perfect example of the scientist who was trying to reconcile his work with his theology. It is worth mentioning that in Robert Boyle's 1680, *Experiments And Notes About The Producibleness Of Chymicall Principles Being Parts Of An Appendix, Designed To Be Added To The Sceptical Chemist*, that in the section on the production of alkalis, he cites King Solomon as a cred-

ible source in the piece on *nitre* and how it reacts with vinegar.[97] He notes that
Proverbs 25 in the Hebrew edition (which he mentions he has translated) the
word *naethar* might refer to the more familiar saltpeter/nitre in Boyle's time.
Proverbs 25:20 reads, "As he that taketh away a garment in cold weather, and as
vinegar upon nitre, so is he that singeth songs to a heavy heart" (King James
Version). However, Boyle goes on to mention that Solomon used something he
identifies as *Egyptian nitre* in his alchemical work. Boyle states that the prophet
Jeremy talks about nitre in Jeremiah 2:22, "For though thou wash` thee with
nitre, and take thee much soap, yet thine iniquity is marked before me, saith the
Lord God"(KJV). However, unlike the usual alkali that is prepared by fire by
alchemists, natural Egyptian nitre was obtained from the evaporation of Nile
River water. Here we see alchemy written about as if it is a natural process, created and given to mankind from God.

Elias Ashmole, a contemporary of the Boyles and a member of the Royal
Society, illustrates this straddling of the line between the scientific exploration
of the universe and superstition.[98] Ashmole was interested in many subjects including astrology and alchemy. He compiled and published the well-known
Theatrum Chemicum Britannicum in 1652. One of Ashmole's friends, Dr. William Currer, was a physician and alchemist who was believed to have achieved
transmutation. Currer was interested in the medicinal use of antimony and arsenic. But C. H. Josten writes that Ashmole also explored magic. Around 1650,
Ashmole claimed that he created a magic sigil that successfully rid his home of
fleas. There is a complex method to creating these magical images, taking into
account astrology and using incantations.[99] Samuel Hartlib heard of the story
from a friend and he believed Ashmole completely. By 1652, Ashmole had included Kabbalistic signs in his engraved sigils. Thus, serious scientists were still
very connected to the spirit world and viewed magic in a very different way than
the modern thinker.

Thus, it is not a great leap to imagine that Katherine was not only familiar
with the occult sciences but perhaps, experimented with the kinds of charms and
magic that to the modern mind seem the antithesis of Christianity. Yet, Christians of the seventeenth century were still people of their time and the world of
magic was not so extraordinary in the way we think of it today. Superstition and
witchcraft were associated with Satan; yet, for many, the occult sciences were
acceptable means for exploring the nature of the universe.

I include this small mystery as an example of how difficult it can be to enter
the world of the women alchemists. I can try my best to place my
preunderstandings and modern filters aside as I sink into a different century;
however, the process is not perfect. There is room for misunderstanding, denial,
and disbelief that the subject of our historical imagination could have behaved in
a way that is contrary to what we seem to have begun to understand. That skepticism is an extremely important check on making wild assumptions with little
evidence. However, there is often a nagging doubt in this type of historical work
that suggests the researcher has only skimmed the surface of the story.

Elizabeth Grey, Countess of Kent: Is it Alchemy or Not?

Lady Elizabeth Talbot (1582-1651), a sister of Alathea Talbot, was educated at home and demonstrated a great interest in languages. Upon marriage to Henry Greaty (1583-1639), she became Countess of Kent. For some reason, Elizabeth would not consummate her marriage for many months and even upon doing so, never became pregnant. She participated in court life, gaining the enviable position of "the queen's first lady of the bedchamber"[100] Henry died in 1639 and Elizabeth was described as being profoundly despondent over his death. Historical records make mention of a Mr. Selden who lived with Elizabeth and Henry in their London home, even after Henry's death. There was a rumor that Selden and Elizabeth married but there is no proof despite the fact that he did become her beneficiary.

I have included Elizabeth in this chapter as an illustration of how some of the recipe books I have examined provide whiffs of alchemy, although not actually discussing it specifically. John Considine mentions that the *Queen's Closet Opened, 1665* and another recipe book attributed to Mr. Kinelm Digby, *The Closet of the Eminently Learned Sir Kenelme Digbie Kt. Opened*, 1669, were written after the fashion of a recipe book attributed to Elizabeth Grey Kent, *A Choice Manuall, or Rare and Select Secrets in Physick and Chyrurgery: Collected, and practiced by the Right Honourable, the Countesse of Kent, late deceased . . . published by W. I. Gent*. In *The Queen's Closet Opened*, there is a reference to "The Receipt of the Lady *Kents* Powder Presented by her Ladyship to the Queen."[101]

Examining the second edition of the *Manuall* published in 1653 is fascinating. The book is quite small, about 2 inches by 4 inches. Being that this is the second edition, the publisher notes that some other experiments and recipes have been added. Samples from the table of contents include directions on how to assist someone suffering with maladies such as: "For an Ach [ache] in the jount . . . For a bruise . . . For burning in the back . . .For a Sore Breast . . . For a stinking breath" and other complaints such as jaundice, bone ache, internal bruising, consumption, cornes and cankers, collick, cuts, dead child in a woman, and falling sickness. Mostly, these recipes call for taking bits of this and that, herbs, wine, etc., and combining them to create the healing powder, drink, or poultice. For example:

> A Plaister for a sore Breast....Take crums of white bread, the tops of Mint chopped small, and boyle them in strong Ale, and make it like a poultesse, and when it is almost boyled, put in the pouder of Ginger, and oil of Thyme, so spread it upon a cloth, it will both draw and heal.[102]

An interesting powder was recommended for black jaundice a bacterial infection of the liver.

For the black Jaundies...Take earth-Wormes, wash them in white Wine, then
dry them, and beat them into pouder, and put to a little Saffron, and drink it in
beer.[103]

There is one recipe that refers to Paracelsus, the well-known alchemist and
physician. It is titled, "To make the best Paracelsus salve."[104] The recipe is
worth quoting as the reader will see that many alchemical terms and operations
are used. It supports the notion that some of what we know as women's alchemy
took place in the kitchen.

> Take of Litharge of Gold and Silver, of each three ounces, and put to it one
> pound and half of good sallade oyle, and as much of Linseed oyle, put it into a
> large earthen vessell well leaded, of the fashion of a milk boul [bowl], or a
> great bason [basin], set it over a gentle fire, and keep it stirring till it begin to
> boyle, then put to it of red Lead, and of *Lapis Calaminaris*, of each half a
> pound, keep it with continuall stirring, and let it boyle two houres, or so long
> till it be something thick, which you may know by dropping a little of it upon a
> cold board or stone, then take a skillet, and put into it a pound of yellow Wax,
> as much black Rosin, half a pound of Gum Sandrach, of yellow Amber, Oliba-
> num, Myrrh, of *Aloes hepatica*, of both the kinds of *Aristolochias* round and
> long, of every of these in fine pouder s(f)earced one ounce, of *Mummia* one
> ounce and a half, of oyle of Bayes half a pound, of oyle of Juniper six ounces,
> dissolve all these together in the aforesaid skillet, and then put them to the for-
> mer Plaister, set it over a gentle fire, and keep it with stirring till it boyle a lit-
> tle.[105]

The directions for the salve continue to describe more mixing of ingredients, boiling of the stuff, dissolution in white wine vinegar, and continued stirring over the fire. Then the directions sound interesting as the reader is instructed to take:

> Of both the Corals red and white, of Mother of Pearl, of Dragons bloud, of *Ter-
> ra lemnia*, of white Vitriol, of each of them once ounce, of *Lapis hematits*, and
> of the Loadstone, of each of them once ounce and a half, of the floures of An-
> timony two drachms, of *Crocus Martis* two drachms, of Camphire once ounce,
> of common Turpentine half a pound, mix all these together, but first let those
> things that are to be poudered be carefully done, and fully searced, then put
> them altogether among the former things, and again set it over the fire with a
> moderate heat and gentle to boyle, till it be in the form of a plaister, the which
> you may know by dropping it on a cold piece of Wood, or Stone, or Iron: you
> must also remember to keep it with continuall stirring from the beginning to the
> ending, when you make it up, let your hands, and the place you roul it on, be
> annointed with the oyle of Saint *Johns* Wort, and of earth Wormes, and Juniper,
> Cammomile and Roses together, wrap it in Parchment, or Leather, and keep it
> for your use. [106]

The use of antimony and Vitriol do not establish the recipes as alchemical. They do, however, illustrate the strangely overlapping worlds of the household and laboratory healing arts.

Comparing the Women Alchemists' Recipe Books with More Traditional Examples of the Genre

I want to include in this book a few examples of recipe books with male authors, in order to illustrate that men have been more easily recognized as alchemists due to their scientific writings; however, many men also compiled recipe books or at least had them published under their name. The reader will see some similarities and some differences between the men's recipe books and those written by the women alchemists. It illustrates how recipe books that are alchemical are different from those that focus solely on household hints.

Examining a variety of alchemical tracts illustrates how these old medical recipes resonate so well as alchemical literature. It may help the reader better understand why emerging historians have made the connection between the two genres. Very little was actually written about technology during the medieval era in Europe; however, in the 1830s, craftsmen such as Charles Eastlake researched the history of painting which led them to collect manuscripts that dealt with pigments. They found directions for making dyes, pigments, colored metal, and other crafts in the Italian *Lucca Manuscript* dating from about the eighth or ninth century.[107] One of the tracts, *De diversis artibus,* was written by Theophilus (a pseudonym) and is unique in that it appears to have been written by the actual practitioner who goes into detail regarding his procedures in the workshop. Thus, the reader is treated to a great amount of detail regarding older techniques that used what we would call alchemical operations.

Mappae Calvicula

Another example of older alchemical/recipe literature is found in a copy of the *Mappae Calvicula*, discovered in the possession of Sir Thomas Phillipps in 1847. There is a record of an additional copy housed in a Benedictine monastery at Lake Constance, believed to date from the year 821-822. Cyril Smith and John Hawthorne translated a copy dating from the twelfth century, as it seemed to be the most intact of the surviving copies.[108] Scholars believe the recipes in the *Mappae Calvicula* go back to antiquity. Although it is a translation, the Smith-Hawthorne version appears to be superior and as the translators note, details are complete regarding medieval technology as well as alchemy in its earliest phases. Research into alchemy and medieval technology is further assisted by paintings, drawings and actual equipment from the labs showing much of the technology.

The first chapters of the *Mappae Calvicula* deal with recipes for the creation of different pigments such as vermilion, azure, and green. The next section addresses the making of gold. One of the recipes follows.

> Take 8 oz. of quicksilver, 4 oz. of gold filings, 5 oz. of Cyprian copper filings, 2 oz. of brass filings, 12 oz. of cleavable alum and the efflorescence of copper which the Greeks call calcantum, [blue vitriol—possibly copper sulfate], 6 oz. of gold-colored orpiment, 12 oz. of *elidrium*. Then mix all the filings with the quicksilver and make it like a wax salve. Put in the *elidrium* and orpiment [arsenic disulphide] at the same time; then add the efflorescence of copper and the alum and put it all in a pan on the coals and cook it lightly, sprinkling over it with your hand an infusion of saffron in vinegar and a little natron [soda] and sprinkle 4 oz. of saffron bit by bit until it dissolves and leave it to intermix. Then when the mixture has coagulated, take if off and you will have gold with increase. Now add also to the above ingredients a little moon-earth, which in Greek is called *Aphroselenos* [i.e., foliated selenite].[109]

Seventy-one recipes follow that include additional ways to make gold as well as directions on how to increase it in mass. Instructions are also given for cooking gold, melting and assaying gold as well as decorating, gilding, and writing with it. The ingredient, *elidrium*, was believed to be swallow-wort or a mineral of some sort. Many of the recipes create gold alloys as opposed to actual transmutation of the proscribed base metal into gold.

Reading the directions for the medical recipes such as those given by Alathea Talbot, Elizabeth Kent, and Katherine Boyle illustrates a basic similar method for either creating a healing plaister or obtaining gold. The operations are nearly the same except that the practitioner did not always refer to the mystical processes typically associated with alchemy. One can imagine that for some practitioners, the process of alchemy explained how the world worked. For others, the *physics* of the work was thought to be understood, but possessing that knowledge was secondary to the product that was being prepared, whether the result be for technology or for healing. The following are examples of male writers in the recipe genre that do not necessarily include directions for transmutation.

Giambattista Della Porta's *Natural Magik*

Giambattista Della Porta, an Italian scholar, lived from approximately 1535-1615. In the preface to *Natural Magik*, he tells the reader that although he is now 50-years-old, the book was begun when he was about 15-years-old. According to him, his recipe book appears to have been popular and was translated into Italian, French, Spanish, and Arabic. Della Porta continues that for the current edition, after 35 years have passed since the first printing, and after he has continued to collect many secrets and recipes, he would like to share them with the reader in this new edition.[110] Della Porta includes items both strange and useful.

He wants to give the reader information while defending himself, asserting that he is not a sorcerer. He warns in strong language that charlatans with their limited knowledge should be avoided. His chapters follow a similar format to other recipe books including chapters on making simples, animal husbandry, and crop planting information. His chapter on cookery contains a helpful remedy for "sobering men up after becoming drunk on wine" as well as "how to tenderize meat."[111]

Della Porta does not shy away from women's health and beauty issues. He gives directions for making perfume and in the chapter on beauty tips, he tackles such topics as keeping breasts from sagging, maintaining healthy teeth and skin, gives notes on caring for childbirth wrinkles, discusses how to blacken eyebrows, and on the topic of hair, the reader learns how to remove it, grow it and color it.

Natural Magik is similar to the women alchemists' work but Della Porta addresses an area that I have not seen in the women's recipe books, providing directions that would be well suited to a modern day version of a manual for intelligence gathering and dispersal. Chapter/Book 16 (he calls them books) is especially interesting in comparison to the traditional recipe books. He titles it: "Wherein are handled secret and undiscovered notes."[112] For example, to make disappearing ink, one may use aqua Fortis (nitric acid). However, if one wishes to make ink that disappears and as well, a courier can eat the paper on which it is written, ink mixed with oil of vitriol seems to do the trick with the added benefit of not killing the messenger. Moreover, preparing ink with orange juice causes it to not only disappear but to also reappear, a method known to even 21st century adolescent sleuths.[113] This seventeenth century book of recipes also explains how one can send a secret message using a servant. After shaving the servant's head, one writes the message on his scalp. Then one must be able to wait for the hair to grow back before sending the servant to the intended recipient of the message. The ink can be an issue due to sweating. However, one can deal with that by tattooing the message instead of just writing it in ink. Finally, the following suggestion reminds us of the place held by four legged creatures in the day. One could place the secret in a dead animal or if preferred, the animal could be alive allowing the recipient to kill it, remove the secret from the bowel, and get the meat as a bonus.

Della Porta's recipe book contains an interesting mix of directions for tempering steel, making seawater potable, exploring the uses of the loadstone, and how to create counterfeit precious stones. His chapter on metallurgy touches upon more traditional alchemical practices by explaining how to turn tin into either silver *or* lead. The reader might wonder at the equal status of the two but may be reminded that lead ammunition was costly and valuable. One last point is that Della Porta includes the directions for making *Artificial Fire*, also known as *Greek Fire*. This was considered especially useful as its composition allowed it to burn upon water, making it ideal for warfare. *Greek Fire* is supposed to have come to Great Britain via the Romans and was made with brimstone, pitch, camphire, saltpeter, lime, and gunpowder.[114] Thus, *Natural Magik* is both similar and dissimilar to the recipe books I have described so far. Yet, I would like to mention a few others that show the recipe books of the women I have presented are truly different from more mainstream recipe books.

Hannah Woolley

Another excellent volume of the recipe genre is Hannah Woolley's 1672, *The Ladies Delight: or, a rich closet of choice experiments & curiosities.* It is worthy of note that even in the title, the emphasis goes beyond cookery and peaks one's interest as to which experiments she is going to discuss. The frontispiece shows a young woman obtaining a liquid from an extraordinary distillation vessel decorated with faces and paws similar to that of a lion's. The vessel straddles a fire that is suspended in a kind of three-legged metal basket. The collecting tube is thin and is attached at the top of the vessel. A coiling metal snake that serves as the on-off spigot surrounds the tube. There appears to be a female servant arranging paraphernalia at a table. Their workspace is truly a kitchen-laboratory considering the collection of bottles, a retort, and vegetables arranged in a basket

Although her focus is on preserving and candying, Hannah Woolley also addresses the usual ailments and concerns such as childbirth, aches, pains, coughs, pimples, and facial care. One of her more interesting cures uses radishes to cure sores that refuse to heal. She states, "To dry and heal up and old sore . . . Take the burned Ashes of a Radish, and strew it upon a Sore, and it will dry and heal."[115]

Diana Astry

Diana Astry's *Recipe Book* includes recipes that represent the traditional types we think of in that genre.[116] Each offering that can be identified by the source is given. For example recipe number five covers "Making Pancakes the Lady Churchell's Way.[117] Cooking puddings, bread, turkey, and eel pie are typical of the contributions. Diana Astry includes directions for preparing healing waters such as plague water—good for measles, pox, and convulsions. "Cinimon water"[118] was thought useful "to be stilled in aches, agues, faintings, & in purging medicines."[119] A cordial for "many distempers"[120] *The Lady Hewet's Cordial Water* (number 274) will also "presarve life a few minutes longer in any dieing pesson if you given them some of it—it will cost 50 shillings a quart to make it."[121] For the most part the recipes are typical of those found in the recipe books of the day and are not alchemical in nature. They often use distillation techniques; however, as I have stated, that alone is not enough for me to place the recipe in the same category as those produced by the women alchemists.

William Lovell

William Lovell's 1661, *The Dukes Desk Newly Broken Up—wherin is discovered divers rare receipts of physick and surgery, good for men, women, and children*[122] is another example of what I am calling a more traditional recipe book. It contains the usual cookery information as well as directions for distilling the "Water of Life"[123] which made use of Rosemary, carnations, hyslop, fennel, sugar, dates, coney, and egg yolks. It was useful for black jaundice and agues. The various medicinal directions are pretty typical of recipe books of the time dealing with aches, pains, headaches, and even a diet remedy using fennel water.[124] He also includes tips for raising livestock. This is not remarkable regarding its lack of alchemy since the *magnum opus* is not standard in all recipe books. Rather, this is a good example of how alchemy was *not* typical of all recipe books and illustrates how the women alchemists' books were different.

Robert Boyle's *Safe Remedies*

I hope the reader will allow one more example because the irony is wonderful. Katherine Boyle's brother, the famous Robert, who is well known for his study of alchemy and physics—as well, wrote his own small recipe book that focused on healing. Boyle states that his *Medical Experiments: or a collection of choice and safe remedies for the most part simple and easily prepared: very useful in families and fitted for the service of country people*[125] was written with the idea that people who lived in rural communities might not have access to a doctor and would find the remedies useful.

Robert tells the reader in his preface that he hopes the book will provide a helpful set of medical recipes, even though he does not consider himself a notable physician. He seems to feel the need to justify his medical work, wondering if the reader will be able to accept him as a reasonable authority considering he has been ill so much of his own life. Robert writes that he attributes much of his illness to his mother dying of consumption at age 42. Also, he mentions that there are some who attribute his poor health to so much time spent studying. He states that he in truth sees his illness as caused by a fall from a horse he had when he was still living in Ireland, "by which I was so bruised, that I feel the bad Effects of it to this day."[126] He goes on to suggest his illness was exacerbated by traveling in inclement weather before he was well, staying in poor rooms in Irish inns, "and the mistake of an unskillful or drunken Guide"[127] that resulted in his getting lost in the mountains where he became ill with fever and dropsy. To make matters worse, when Robert arrived in London, there was a sickness going around the city that caused him to become even more ill than he had already been.

Robert recovered but had many health problems that he lists such as poor eyesight that he attributes to reading too many Hebrew books while sick and other "Eastern characters."[128] He experienced difficulty in his limbs that left him virtually immobile for months. Palsy and kidney stones are also mentioned in his list of complaints. The point he hopes to make for the reader is that despite his illnesses, or probably because of them, he has learned many useful medical directions that he can explain in simple terms that are clear and easy to use. His own health is not a reflection of the effectiveness of the remedies he has included in this book.

Robert's medical recipes can be found in the Boyle Papers[129] and follow a similar pattern to the ones written by the women in this chapter. His alchemical symbols and directions are scattered throughout his directions and one gets a sense that for some authors, alchemy could be presented in a rather mundane style without green dragons, kings and queens. Is this practical vision of alchemy less unusual than the more esoteric writing? I suggest that they are just different in style and intent. Paracelsus wrote extensively about using alchemical operations to create healing elixirs. The recipe books discussed in this chapter are not so elaborate or grand but the fact that they were handed down through families illustrates their usefulness and value to the writers.

Notes

1. Walter Endrei, *Old Chemical Symbols* (Budapest: Hungarian Academy of Arts and Crafts, 1974), 5.

2. Wolfgang Schneider in *Science, Medicine, and Society in the Renaissance; Essays to Honor Walter Pagel,* Allen Debus, ed.,volume 1:141-150.

3. Marie Boas, *Robert Boyle and Seventeenth-Century Chemistry* (Cambridge: Cambridge University Press, 1958).

4. Lynette Hunter and Sarah Hutton, eds., *Women, Science and Medicine 1500 – 1700: Mothers and Sisters of the Royal Society* (Gloucestershire: Sutton, 1997).

5. Giambattista Della Porta, *Natural Magick by John Baptista Porta, A Neapolitane: in Twenty Books* (London: Thomas Young and Samuel Speed, 1658).

6. Robert Boyle, *Medical Experiments: or a collection of choice and safe remedies for the most part simple and easily prepared: very useful in families and fitted for the service of country people, 4th edition* (London: Sam Smith, at the Prince's Arms in St. Paul's Churchyard, 1703).

7. Lucia L. Tosi, "Marie Meurdrac: Paracelsian Chemist and Feminist." *Ambix,* 48 (July 2001): 69-82. See also Jean Jacques (trans.). According to some historians, the first edition was published in 1656; however, the 1666 edition is also referenced by others.

8. *La Chymie charitable et facile, en faveur des Dames Charitable* [*Easy Chemistry for Women*], 1666 Jean Jacques edition. Jacques uses an edition of *La Chymie* that had been published in 1856 by a scholar he identifies as Moreau who had researched the Meurdrac family extensively. Moreau places Marie's birth quite a bit before Catherine's by means of the records of a godmother by that name. However, remarks made by the publisher of the third edition of *La Chymie* suggest that Marie died at a young age. It is more probable that the Marie Meurdrac who published *La Chymie*, was named after the godmother mentioned earlier and may have been born closer to her sister, anywhere from 1600-1612. Yet, that does not account for the comment made by the publisher regarding her untimely death.

9. Unpublished translation, Christy A. Roach, 2009.

10. Lucia Tosi (2001); Ogilvie and Harvey (2000).

11. Marie Meurdrac, *La Chymie charitable et facile, en faveur des Dames Charitable* [*Easy Chemistry for Women*], 1666.

12. Tosi, 71.

13. Marie Meurdrac, *La Chymie charitable et facile, en faveur des Dames Charitable, 1666* [*Easy Chemistry for Women*] (Paris: CNRS Editions, 1999), 17. (unpublished translation by Christy Ann Roach)

14. Meurdrac, 21.

15. Meurdrac, 175-176.

16. Tosi (2001).

17. Meurdrac, 115.

18. Meurdrac, 129.

19. Meurdrac, 131.

20. William Eamon, *The Professor of Secrets: Mystery, Medicine, and Alchemy in Renaissance Italy* (Washington, D.C.: National Geographic, 2010).

21. Isabella Cortese, I secreti *ne'qvali si contengono cose minerali, medicinali, arteficiose* (Venice: Appresso Giouanni Bariletto, 1565). Replica printed by Book Renaissance, n.d.

22. Eamon, 117.

23. Eamon, 117.

24. Cortese, 1.

25. Cortese, 1.

26 Cortese, 3.

27. Cortese, 28.

28. Cortese, 28.

29. Cortese, 50.

30. Hunter (1997).

31. Matthew and Harrison (2004).

32. Matthew and Harrison, 439-440.

33. Matthew and Harrison, 439-447.

34. Alathea Talbot, *Natura Exenterata: Or Nature Unbowelled.* (London: H. Twiford, 1655).

35. Talbot, 375.

36. Talbot, 375.

37. Martinus Rulandus, *Alchemical Lexicon* (Frankfurt: Zachariah Palthenus, 1612), 273.

38. Talbot, 376.

39. Talbot, 376.

40. Talbot, 380

41. Talbot, 380.

42. *The Queen's Closet,* London: Obdiah Blagrave at the Sign of the Black Bear in St. Pauls Church-yard, 1683.

43. *The Queen's Closet* (1683).

44. R. H., Fritze and W. Robison, eds., *Historical Dictionary of Stuart England, 1603–1689* (Westport, CT: Greenwood Press, 1996), 469-470.

45. Fritze and Robinson, 474.

46. Mendelson, 62.

47. Hutton (1992); Matthew and Harrison (2004).

48. Mendelson (1987).

49. Charlotte Fell Smith, *Mary Rich, Countess of Warwick (1625 - 1678): Her Family & Friends* (London: Longmans, Green, and Co., 1901), 37.

50. Mendelson, 67.

51. Maddison, 4.

52. Note regarding a payment made by the Earl of Cork, Katherine's father, on April 8, 1630, "hath my bond for payment of my daughter Katherynes marriadg portion, being three Thowsand pounds ster [sterling]." *The Complete Peerage or a History of the House of Lords and All Its Members from the Earliest Times*, Volume 10 (London: St. Catherine Press, 1945), 732.

53. Matthew and Harrison, 574.

54. Fell Smith, 37.

55. Mendelson (1987).

56. Mendelson, 9.

57. Fell Smith, 313-314.

58. R.E.W. Maddison, *The Life of the Honourable Robert Boyle F.R.S.* (London: Taylor and Francis, 1969).

59. Fell Smith, 1901.

60. Henry W. Robinson and Walter Adams, *The Diary of Robert Hooke, M.A., M.D., F.R.S. 1672-1680* (London: Taylor and Francis, 1935), 308.

61. Robinson and Adams, 334.

62. Robinson and Adams, 335.

63. Robinson and Adams, 335.

64. Robinson and Adams, 364.

65. Robinson and Adams, 365.

66. William Robertson, *The First Gate, or, The Outward Door to the Holy Tongue, opened in English* (London: Evan Tyler, for Humphrey Robinson, at the three Pigeons in St. Paul's Churchyard, 1654).

67. Robertson, A3.

68. William Robertson, *The Second Gate, or, The Inner Door to the Holy Tongue* (London: Evan Tyler, for Humphrey Robinson, at the three Pigeons in St. Paul's Churchyard, 1655).

69. Marie Boas, *Robert Boyle and Seventeenth-Century Chemistry* (Cambridge: Cambridge University Press, 1958).

70. Matthew and Harrison (2004).

71. *WMS 1340*, Wellcome Library, London. Boyle Family. *Collection of 712 medical receipts, with some cookery receipts, mainly written by two hands; circa 1675-circa 1710*. Also MS Sloane 1367, British Library, London.

72. Hunter and Principe, 50.

73. Carl Zimmer, *Soul Made Flesh: The Discovery of the Brain—and How It Changed the World* (New York: Free Press, 2004).

74. Hunter, 188.

75. Fritze and Robison, 473.

76. *MS Sloane 1367*, 2.

77. *MS Sloane 1367*, 93-94.

78. *MS Sloane* 1367, 93-94.

79. *WMS 1340*, 40.

80. *MS Sloane 1367*, 48.

81. Rulandus, 283.

82. *WMS 1340*, 299.

83. Barbara Walker, *The Women's Dictionary of Symbols and Sacred Objects* (San Francisco: Harper,1988), 7 and 49.

84. Carl G. Liungman, *Symbols: Encyclopedia of Western Signs and Ideograms* (Stockholm: HME Publishing, 1995/2004), 283.

85. Harold Bayley, *The Lost Language of Symbolism,* Volumes I-II (London: Williams and Norgate, 1912), 60.

86. Liungman, 252.

87. Liungman, 45.

88. Liungman, 249.

89. Fred Gettings, *Dictionary of Occult, Hermetic and Alchemical Sigils* (London: Routledge & Kegan Paul, 1981).

90. Gettings, 198.

91. Walker (1988).

92. http://www.bbk.ac.uk/boyle/researchers/researchers_area_homepage.htm

93. *WMS 1340*.

94. Wouter J. Hanegraaff, *Dictionary of Gnosis and Western Esotericism*, Volumes 1 and 2 (Leiden: Brill, 2005), 201.

95. William Kingsland, *The Physics of the Secret Doctrine* (London: Theosophical Publishing Society, 1910).

96. Mitchell Salem Fisher, *Robert Boyle: Devout Naturalist, A Study In Science And Religion In The Seventeenth Century* (Philadelphia: Oshiver Studio Press, 1945).

97. Robert Boyle, *Experiments and notes about the producibleness of chymicall principles being parts of an appendix, designed to be added to the Sceptical Chemist* (Oxford: H. Hall for Ric. Davis,1680), 30-32.

98. C. H. Josten , ed., *Elias Ashmole (1617-1692): His autobiographical and historical notes, his correspondence, and other contemporary sources relating to his life and work* (Oxford: Clarendon Press, 1966).

99. Josten, 82.

100. Considine, 833.

101. Considine, 833.

102. Kent, 38.

103. Kent, 8.

104. Kent, 157.

105. Kent,157-161.

106. Kent, 157-161.

107. Cyril Stanley Smith and John G. Hawthorne, "Mappae Clavicula: An Annotated Translation Based on a Collation of the Sélestat and Phillips-Corning Manuscripts with Reproductions of the Two Manuscripts." *Transactions of the American Philosophical Society*, 64 (number 4) (Philadelphia: American Philosophical Society, 1974): 3-128.

108. Smith et. al, 28-29.

109. Smith et. al, 28-29.

110. Giambattista Della Porta, *Natural Magick by John Baptista Porta, A Neapolitane: in Twenty Books* (London: Thomas Young and Samuel Speed, 1658).

111. Della Porta, 340.

112. Della Porta, 340.

113. Della Porta, 350.

114. Della Porta, 291.

115. Hannah Woolley, *The Ladies Delight: or, a rich closet of choice experiments and curiosities* (London: I. Pilbourn, 1672), 290.

116. Bette Stitt , ed., "Diana Astry's Recipe Book c. 1700." *Bedfordshire Historical Record Society,* Vol. 37 (Luton, Bedfordshire: Society at Streatley, 1955): 83-168.

117. Stitt, 90.

118. Stitt, 156

119. Stitt, 156.

120. Stitt, 144.

121. Stitt, 144.

122. William Lovell, *The Dukes Desk Newly Broken Up—-wherin is discovered divers rare receipts of physick and surgery, good for men, women, and children* (Lovell, William. London: John Garway, 1661).

123. Lovell, 72-74.

124. Lovell, 93.

125. Robert Boyle, *Medical Experiments: Or a Collection of Choice and Safe Remedies for the Most Part Simple and Easily Prepared: Very Useful in Families and Fitted for the Service of Country People,* 4th edition (London: Sam Smith, at the Prince's Arms in St. Paul's Churchyard, 1703).

126. Boyle, A4-5.

127. Boyle, A5.

128. Boyle, A5.

129. See the web site for the Boyle Papers:
http://www.bbk.ac.uk/boyle/boyle_papers/boylepapers_index.htm

Chapter 6

The Natural Philosophers

"Of Mlle. Germaine Tailleferre one can only repeat Dr. Johnson's dictum concerning a woman preacher, transposed into terms of music. 'Sir, a woman's composing is like a dog's walking on his hind legs. It is not done well, but you are surprised to find it done at all.'"[1]

The above quote was noted by Virginia Woolf regarding a comment made in 1928 about the ability of women to compose music. This rather shocking critique, at least to my more modern sensibility, is not so different from remarks made about many of the women alchemists. Countless women have focused their research on the physical world from antiquity through the 21st century. Several women alchemists sought to explore the nature of the universe using the same scientific paradigm as Isaac Newton and Robert Boyle. It must be remembered that the core concepts regarding matter centered on alchemical processes as the theory for understanding natural phenomena. Achieving transmutation could provide the needed proof that the theory was correct. The women who are the focus of this book were informed by alchemical theory in the same way as their male counterparts. Sometimes their work was tolerated and even admired. Other women had to feel the sting of a society that was very unkind to women who did not limit their interests and studies to subjects deemed acceptable by their peers. The following women alchemists illustrate both attitudes.

Margaret and Anne Clifford: The Alchemist and the Diarist

I debated whether to include Margaret Clifford in the previous chapter on healers in light of her recipe book that contains similar directions for creating medicines. However, one difference in her work is that her surviving recipes seem to me to focus on notes made to herself or perhaps for a colleague, as opposed to recipes to be disseminated to friends and family. The work, *Receipts of Lady Margaret Wife of George, 3rd Earl of Cumberland for Elixirs, Tinctures, Electuaries, Cordials, Waters, etc., MS circa 1550 with Her Annotations* can be found in the Cumbrian Archive in the Cumbrian Record Office in Kendal.[2] The 138 folios were bound in 1920 which may explain why they appear quite different from those bound near the time of the writing such as *Natura Exenterata* (1655).

Another difference in Margaret's recipes is their ease of decipherability, or perhaps it would be more accurate to say their dis-ease. Whereas the previous recipe books I have discussed are fairly simple to follow, Margaret's *Receipts* would have been impossible for a reader who did not have a complex grounding in alchemy, Latin, Italian, and her particular notation system. Both Lady Katherine Ranelagh and Alathea Talbot included tables in their recipe books that defined their notation systems making it clear to the reader what ingredient was needed or used in the work. Furthermore, one can read their narratives fairly easily as they are straightforward. I think what makes Margaret Clifford's *Receipts*, also known as *The Margaret Manuscript*, so difficult to decipher is the directions are interspersed with abbreviations, symbols for the elements, and alchemical terms from both Latin and Italian. The handwriting seems to be from at least two different people and there is debate whether any of these are actually Margaret's. In modern terms, we would say her book is not user friendly. I will go into this in more detail regarding her recipes.

Biographical Notes about Mother and Daughter

Margaret Clifford, Countess of Cumberland (1560-1616)

Margaret Clifford warrants a biographical notation in *The Oxford Dictionary of National Biography*,[3] however; her daughter, Anne Clifford Sackville provides wonderful depth about her mother in her own personal diary.[4] I will use both sources as well as notes from a family biographer to introduce the reader to these remarkable women.

Margaret Russell Clifford was born on July 7, 1560 in Exeter. Her father, Francis Russell, was 2nd Earl of Bedford. Her mother, also Margaret, was the daughter of Sir John St John of Bletsoe. Margaret's mother died when she was 2 years-old. Margaret was then raised by her aunt, Mrs. Elmes of Lillford. Margaret seems to have been given a substantial education and revealed an interest in science that would become significant as an adult.

Margaret married George Clifford, 3rd Earl of Cumberland on June 24, 1577. George was a ward of her father and was described as a rake and a womanizer. They had been married about nine years when George was said to have begun to travel extensively in the service of Queen Elizabeth I. He was known to be an adventurer or in more specific terms, a rather successful brigand for the Queen. The biographer, V. Sackville-West quotes from Lady Anne's diary and adds a few notes of his own, regarding George Clifford:

> 'These marine adventures being neither his profession nor yet urged by necessity thereunto, yet such was his lordship's natural inclination.' On eleven sepa-

rate occasions his lordship's natural inclination prompted him to fit out a fleet (at his own expense) and to set sail for the River Plate, for Costa Rica, the Azores, the Canaries, or the coasts of Spain and Portugal, with the frank intention of plundering such towns as he could reduce and of taking captive such foreign merchantmen as he should encounter. He was wounded in battle, he nearly died of thirst, he was all but wrecked, but, undeterred, after each return to England he at once set about organising a fresh expedition, 'his spark of adventure further kindled and inflamed by former disasters.'[5]

George and Margaret grew apart, separating in the year 1600 with a level of animosity that is quite obvious when Anne writes about their need to meet regarding a land dispute. "Sometimes my Mother and he did meet, when their countenance did show the dislike they had one of the other, yet he would speak to me in a slight fashion and give me his blessing."[6] There was some small reconciliation as can happen with old age as George did repent before he died of "all the wrongs he had done her"[7] which included adultery. Margaret's two sons died leaving Anne Clifford, the youngest, to be the sole survivor and Margaret's focus of attention.

Margaret Clifford is described as being "exceedingly pious ... a zealous puritan."[8] She is further described by Sackville-West (a descendent so perhaps a little romanticized?) as, "Grey-eyed, pensive, and delicate, consoled only by religion and charity, a distiller of medicinal waters, a dabbler in alchemy, an expert in the properties of plants, flowers and herbs, and in her manner civil and courteous to all sorts of people."[9] Margaret Clifford founded an almshouse for poor widows in 1593 called Beamsley Hospital.

Margaret's intellectual interests included literature, poetry, science, and business. Her scientific pursuits encompassed alchemy, medicine and experimentation with smelting iron using coal. She also invested in mining. Her mining partner was a fellow, Richard Cavendish, who died in 1601, a year after Margaret's separation from her husband. She had a memorial erected for him at Hornsey Church, Middlesex suggesting he must have meant a lot to her.

Anne Clifford, Countess of Pembroke, Dorset, Montgomery (1590-1676)

Anne Clifford, Margaret's daughter, was born on January 30, 1590 at Skipton Castle in West Riding, Yorkshire. Anne received her education from both her mother and her maternal relatives (the Russells) who are all described as "intellectually gifted."[10] She also had a tutor (Samuel Daniel) and governess (Anne Taylour) who had a hand in her education. Mrs. Taylour is thought to have been the wife to alchemist Christopher Taylour who worked with Anne's mother, Margaret, and is a probable contributor to the recipe book.

Books were Anne's refuge in troubled times. She was exceptionally well read in a variety of disciplines. The family was also well connected to the royal household. Queen Elizabeth I died when Anne was 13-years-old. Her mother, Margaret, sat with the body for a few nights but Anne writes that she was considered too young to join her. Anne's family continued to be a part of the nobility who surrounded King James I and his Queen, Anne of Denmark. Anne's parents attended the coronation but Anne was not allowed to be present due to her mother's fear of plague which was going around London at the time.

Repeating the marital story of her mother, Anne married Richard Sackville, Lord Buckhurst, and 3rd Earl of Dorset on February 25, 1609. Anne's husband was another womanizer and greedy as well. Richard Sackville is described as an extravagant host; he "squandered his [money] on clothes, hospitality, women, gaming—the first wastrel of a family which was to become notorious for its prodigality."[11] They had five children; only the two daughters lived whereas her three sons died. They were named Margaret and Isabella.

Anne's father had died when she was 15 years-old and his will became the source of a long drawn-out law suit. Basically, George Clifford left his estate to his brother, Sir Francis Clifford. It would revert to Anne if Francis died without a male heir. Where George really stumbled is that many years previously during the time of Edward II, his estate had been entailed to his child regardless of its sex. Anne was therefore the rightful heir. Anne's diary is very much focused on the lawsuit and quest to return her land. A brief narrative concerning Anne's fight for her land will deepen our understanding of the types of women both she and her mother came to be. Anne had to face a great deal of pressure regarding her lawsuit. She notes on Feb. 16, 1616:

> My Coz. [cousin] Russell came to me the same day and chid me and told me of all my faults and errors in this business; he made me weep bitterly; then I spoke a prayer of Owens and went to see my Lady Wotten at Whitehall where we walked 5 or 6 turns but spoke nothing of this business though her heart and mine were full of it – from hence I went to the Abbey at Westminster where I saw the Queen of Scots, her tomb and all the other tombs, and came home by water where I took an extreme cold.[12]

The Archbishop of Canterbury joined the pressure, arguing with Anne for one hour and a half, but Anne stayed firm and wanted to discuss the issues with her mother. Anne states that the group of men with the Archbishop vacillated between threats and flattery but finally agreed to Anne giving an answer after seeing her mother. Yet, it only delayed the answer of no; Margaret and Anne were resolved to maintain Anne's right to her father's lands. Although Anne lived in southern England, her family's land was in the north near the Borders and she loved it well.

Her husband, Richard Sackville, was furious. He tried to manipulate her by insisting that their daughter, Margaret, referred to as "the Child"[13] in the diary, travel to him in London as he knew this would upset Anne greatly. Anne decided not to fight him on this, worrying that it would upset Margaret as much as anything. Richard upped the ante by then writing that he should keep their daughter and Anne would not see her again. In May 1616, Anne wrote:

> All this time my Lord was in London where he had all and infinite great resort coming to him. He went much abroad to Cocking, to Bowling Alleys, to Plays and Horse Races, and commended by all the world. I stayed in the country having many times a sorrowful and heavy heart, and being condemned by most folks because I would not consent to the agreements, so as I may truly say, I am like an owl in the desert.[14]

At the same time, Anne's mother, Margaret, was very ill and died on May 24, 1616. For a time, there was some reconciliation between Richard and Anne, perhaps brought about by Margaret's death.

By December 1616-January 1617, Anne was living in London with Richard and their daughter. King James I was now brought into the lawsuit, trying to get the inheritance business settled between the families. The truce between Anne and Richard was short lived. She mentions that she and Richard had "a falling out about the land"[15] on January 8. He stormed off to London and Anne was summoned before the King. He asked the couple to "put the whole matter wholly into his hands"[16] and to come to terms with each other. Richard agreed, probably having a positive outcome already planned, but Anne writes:

> I beseech'd His Majesty to pardon me for that I would never part from Westmoreland while I lived upon any condition whatsoever. Sometimes he used fair means and persuasions and sometimes foul means but I was resolved before so as nothing would move me.[17]

One has to admire her resolve and bravery. Days later, Anne was called back to the King's Drawing Chamber where she was confronted by her Uncle Francis, Cousin Henry Clifford, various Lords and lawyers. The King asked if they would again let him decide how to move forward but Anne, alone, said no. The King was furious; Anne was escorted out of the Chambers by Sir John Digby in the hopes that he could persuade her but she remained adamant. It also appears that Richard may have had some regrets about bringing the King into the argument, worrying that the King would do something drastic, but Anne writes that she counts God on her side and she is thankful. Not many men or women would have survived this degree of opposition.

The King was so angry that he threatened that Anne's lands would be taken from her regardless of her wishes but she put her faith in God and hoped for the best. On April 17, 1617 Richard told Anne he would not bother her anymore about her lands, seeing how firm she was; but within two days he was at it again and furious at her stalwart determination. She was informed in May that her lands were to be given to her Uncle. Anne was bereft and spent the rest of the year as she records feeling sad and melancholy, but also unwilling to give up her suit.

Richard died in 1624, only 35-years-old. Richard's brother inherited from him causing Anne and the two girls to move to another home. At the same time, she and the girls caught and recovered from smallpox. There was some scarring on Anne's face which caused her to resolve to stay unmarried. Yet, she did re-marry, Philip Herbert, which was yet another marital mistake. Anne made a seemingly fortuitous marriage to Philip Herbert, Earl of Montgomery and 4th Earl of Pembroke. Although the marriage brought Anne a measure of financial security as well as access to the Crown, by around 1634 Anne was sent away from London. The spouses' argument focused on Philip's marital wandering, again repeating the story of marital difficulty experienced by both Anne and her mother. Philip is described as 'violent and contemptible, indeed almost crazy, contemptuous of all culture, careless and cross, false, cruel and cowardly.'[18] Anne and Philip remained married but lived apart.

It took 38 years for Anne to triumph although it was mainly due to the deaths of both her uncle and his son who left no heir. She came into her inheritance when she was 53-years-old and went on to rebuild her castles, heal the land and dote on seventeen grandchildren and nineteen great-grandchildren until she was age 86. There is a story that was told about Anne that was consistent with the strong willed woman she grew to be. A tenant of Lady Anne's owed her one hen as part of his yearly rent. She took him to court spending 400 £, a huge amount of money at that time. But she did win her case. It was reported that, "She asked her adversary to dinner, and shared with the disputed bird as the pièce de rèsistance of the meal."[19] That says much about Lady Anne!

By all accounts, Anne was schooled in alchemy by her mother, Margaret. Anne does not discuss this in her diary which creates some hesitation in calling her an alchemist. However, she was well aware of the science and the work her mother did, providing clues to their level of involvement. I have a favorite quote from Lady Anne's diary that to me describes her relationship to the natural world. She was up late star-gazing and was caught by the beauty of the Dog Star also known as Sirius or as she calls it, the Fire Dog in the constellation, Canis Major. The date is October 18, 1619 and she records: "Upon the 18th at night the Fire Dog play'd with fire, so as I took cold with standing in the window."[20]

The Margaret Manuscript

Penny Bayer completed a wonderful analysis of *Lady Margaret Clifford's Alchemical Receipt Book*. Although the book is housed at the Cumbrian Record Office, I was able to secure a copy on a CD which I will discuss shortly. First I would like to summarize Bayer's analysis which is really in depth and well researched.[21]

This particular recipe book is often called *The Margaret Manuscript*. It was bound in 1920 and the cover reads: *Receipts of Lady Margaret Wife of George, 3r Earl of Cumberland for Elixirs, Tinctures, Electuaries, Cordials, Waters, etc., MS circa (1550) with Her Annotations.*[22] Bayer notes that the handwriting in the book is most likely that of two different people, one being Christopher Taylour who was a known alchemist and a probable acquaintance. As is so often the case in writing about women alchemists, historians need to prove that the recipe books are evidence of their alchemical practice. Bayer notes that Margaret's daughter, Anne, commissioned a portrait of her mother to be painted in 1646 that provides further evidence than the book. Anne ensured that her mother was painted holding a copy of the *Psalms of David* while behind her, sitting on a shelf is a book titled, "*a written hand Booke of Alkumiste Apstracions of Distillation & Excellent Medicines*"[23] as well as the *Holy Bible* and a translation of a writing by Seneca. Anne also wrote in the family's *Great Book of Records of the Cliffords* regarding her mother:

> [Margaret] was a lover of the Study and practice of Alchimy, by which she found out excellent Medicines, that did much good to many; she delighted in the Distilling of waters, and other Chymical extractions, for she had some knowledge in most kinds of Minerals, herbs, flowers and plants.[24]

Finally, Bayer cites a letter written to Margaret by a counselor to Elizabeth I, Lord Willoughby, in which he admires "her work as an alchemist."[25] Willoughby was well versed in alchemy, having close ties to the John Dee Circle. Bayer cites a recipe, "To make a golden or silver tree to growe in a glasse from mercury"[26] that is attributed to a close associate of both John Dee and his assistant, Edward Kelley. The Duke of Brunswick (probably Heinrich Julius, the son of the Duke who put Anna Maria Zieglerin to death as described in chapter four) is given credit for the contribution. Elizabeth I was very interested in the possibility of transmutation. More will be said about this in the section about Lady Mary Sidney. Lord Willoughby's letter is included with the *Margaret Manuscript* papers and reads:

> My honorable and dere honored Lady:
> I can not say how mutch I continue still to be bound to you . . . . But noble phylosophying Lady you have learned the art of Separation, to draw ye Spirit from ye body to add it to it agayne [again], things dead to live, livinge things apparent dead, yet having in concealed beinge a multiplied life. Theses are not natures effects but wisdom's works, of a lytle red Sand to make a Great deale of gold as in Ecclesiastes ye 8. So did Hermes, so did Salomon [King Solomon], so did Ripley and so did Kelley[Edward]. And now come my Lady of Cumberland knowinge how of dissolved putrefied bodies no good can be looked for wthhout Sublimatinge. She brings into ye Alembic head of fine conceite, than digests, than revives it, and lastly projects that rectefied oyle of Gratiounes upon an old bowed plate of Suturne, and by her artifice makes ye bowed mans selfe beleave he is not metall mutch inferiour to his grandchild Sol [gold].[27]

It is remarkable that Willoughby refers to transmutation of metal to gold, separate from any reference to healing medicines.

The notes in the margins of the recipe book were believed by an archivist for the Cumbria County Council to be Margaret's. However, Penny Bayer states that the handwriting in two different hands does not match Margaret's writing in extant letters. More research is needed to clarify this issue; however, it is not that uncommon for these books to have multiple hands. Bayer believes one of the hands is from Christopher Taylour who was an alchemist and his relationship to Margaret is one of mostly conjecture at this point. Bayer cites page 35 from *The Margaret Manuscript* as an example of the work from Christopher Taylour due to the signature, C.T. and the date, 1598.

As I mentioned in the beginning of this section, *The Margaret Manuscript* is not easy to decipher. Between the unclear hand, the mixing of Latin, English and Italian, and the profuse use of alchemical nomenclature, it is difficult to follow the recipes in the same way that is possible with those such as Lady Katherine Ranelagh's.

There is no doubt that the recipe book is alchemical in nature. It contains directions for healing medicines as well as the best way to work with minerals. The notes in the margins are especially useful as they tell the reader whether the recipe is an effective one or not, and furthermore, add clarifications to the directions. In discussing specific recipes, when the symbol is used with no explanation, I have indicated the symbol as well as noted the name that was typical for the time. For example the title for one offering is, "A Mercuriall water to dissolve [symbol for gold—☉] and to make an Elixer thereof given by Sr Ed Kellye unto the Lorde Willobye att his departure from him at his house neer Prage."[28] The alchemical symbol for gold is fairly consistent across time and cultures. Although the word gold is not named specifically in the recipe, I believe it is safe to read the symbol as gold. I have also attempted to retain the punctuation.

The Margaret Manuscript also gives a glimpse into Margaret Clifford's circle of acquaintances by referencing the person who either created the recipe or passed it along. Edward Kelley was the associate of the famed alchemist, John Dee, who worked for Queen Elizabeth I as mentioned earlier. Another recipe is noted as, "An excellent preparation of Silver for paynes in the Heade & other Diseases called by my L. Willoughby the vitriol of [symbol for possibly argentum or silver—☾. The symbol for silver usually points toward the left—☽]."[29] Raymond Lull, a famed Spanish alchemist, is also referenced more than once with a suggestion that the reader should study his work. Other notable alchemical names are likewise discussed such as Paracelsus, Petrus Bonus, and Fodoci Greberi who may be the Arabian writer, Geber.

For the recipe, "To prepare the Essence or Tincture of Antimonye" there is a side note— "This maner of practise to make the Essence or Tincture of [symbol cross over circle—antimony ♀] is not good and did prove of noe worth."[30] Antimony was often used by alchemists in healing although there were some who condemned it for its potentially poisonous side effects. A physician, Theodor Kerckring (1640-1693) wrote about Basil Valentine's work with antimony and its uses to heal disease.[31] There is some foundation for the concern about poison since Martinus Rulandus (1612) writes "antimonium or alkofol—is a stone from a lead vein . . . . It is of two kinds. One is the ordinary black species of Saturn, and is called Magnesia, Bismuth, Contersin . . . has the virtues of burnt Lead."[32] Modern use of antimony is found in fire retardants and microelectronics.

Throughout *The Margaret Manuscript* there are pages of directions using typical alchemical terms such as heating in the athanor (oven), symbols for the terms, day and night, and as well, a symbol that in Margaret Cavendish's *Natura Exenterata* means, "to purify."[33] (See Figure 6.1)

Figure 6.1: Samples of General Alchemical Notation used in *The Margaret Manuscript.*

One recipe concerns an everlasting lamp of some sort. "Note./ Reade more of this Spirite, to make a perpetuall burninge Lampe in a booke called Aureum Vellus, printed in Ducheit wth picters./"[34] There is also some suggestion that Margaret worked with transmutation; "To transmute [symbol for silver—☽] into [symbol for gold—☉] proved true by Mr G/Cromptonn."[35] A few of the notes seem to be signed with a letter "M" which may be the source of thinking that the annotations were from her. One example reads: "Note./ Studdye well my ecphrasticall Comentarie upon Geber, of his books of Alchymie: written by me, And is amongst my books, you shall fynde good doctrine therin."[36] "M" is the signature and as well, follows a note cited as Christopher Taylour's. "Note there is another written booke in frenche of Rovilliascos wch is of preparinge of Mercuris waters or his Aqua vite. C.T. / M."[37]

The Margaret Manuscript also contains an interesting recipe titled, "The Heaven of the Philosophers./" where use is made of the water bath or "Balneo Mari to putrifie."[38] This elixir can be traced to another alchemist, William Baro, also known as Urbigerus (1616-1691) who published the 1690 *Aphorismi Urbigerani, or, Certain rules clearly demonstrating the three infallible ways of preparing the grand elixir, or circulatum majus of the philosophers : discovering the secret of secrets, and detecting the errors of vulgar chymists in their operat.*[39] "46. When our Chaos or Celestial water has purified itself from its own gross and palpable body, it is called the Heaven of the Philosophers, but the palpable body is the Earth, which is void, empty, and dark."[40] The actual elements found in "The Heaven of the Philosophers" are rather unclear but the substance was very important to certain alchemists.

The most *alchemical* sounding language is found in the two following notes in the recipe book. The term *Green Lion* was used by alchemists to refer to separating out the philosophical mercury (not the element mercury) from the gross matter, an early stage in producing the *Philosopher's Stone.*[41] The *Red Lion* is another stage in the process whereas the *White Eagle* may refer to the sublimation stage where the liquid is vaporized.

> Note./ But knowe yt the Mercurye drawne out of the read Leade of ye phersor ye stone it selfee reduced againe into Mercurye vive, is the true [symbol] most pfect oylye Mercuri, then who doth make the worcke very shorte and speadye [short and speedy]. And this is the oylye Axoc drawne out of the red calx of ye philosophers, or any of these red duste wch is called Sol precipitate: And also the Blowde [Blood]of the Reade Lion./"
>
> Note./ And alsoe knowe that the Mercury vive of vulgar [symbol for gold] is called the White Eagle./ And some tyme the Grene Lyon.///"[42]

Margaret and Anne Clifford are examples of women alchemists whose link to the work is not always overt. Their association with known alchemical practitioners and comments recorded by family provide evidence that can be challenged; however, looking at it simply, their work in alchemy appears to have been acknowledged and accepted. Rebecca La Roche suggests that the *Clifford Family Great Painting* depicts Anne with a book on herbals that she may have authored herself.[43] She argues that evidence for the work women did in numerous areas is not only found in recipes, memoirs, and diaries but in "print manuals, inscriptions, memorials, wills, and historical records."[44]

The women alchemists, more often than not, failed to refer to themselves as *alchemists*. Lady Margaret Hoby serves as an example of the challenge of defining the work of women alchemists. Lady Margaret Hoby became well known for her medical skills but she represents the author of the types of recipe books that I am not including as evidence for alchemical practice in the strict sense of the definition.

Margaret Hoby's focus is on healing but there is no evidence that she was thinking beyond that. She writes profusely about her medical art but her philosophy focuses on God, not natural philosophy as a science. However, she did note that she had distilled aqua vitae on July 20, 1601.[45] Johanna Moody suggests that aqua vitae referred to an ale or whiskey; however, in alchemy, aqua vitae is an important ingredient in extracting the essence from the plant or mineral. Martinus Rulandus describes aqua vitae as, "Mercury, but the term is sometimes used for distilled Wine, and for various Waters mixed with distilled Wine. That is consistent with the making of tinctures where the plant is soaked in a liquid with a high concentration of alcohol, such as in wine. It is obvious from Margaret Hoby's diary that she knows about *chymstry*, to use the older term, but

I do not feel I have the needed evidence to place her in the group of women alchemists I am discussing in this book. She was well read in herbals and seems to have been sought out for her medical skill as well. Mary Sidney Herbert, the next woman I will discuss falls more into the "guilt by association" category.

Mary Sidney Herbert, Countess of Pembroke (1561-1621): In the Sidney Circle

Mary Sidney Herbert was actually a contemporary of Margaret Clifford, Anne's mother. I apologize to the reader ahead of time if there is some difficulty in keeping these different relationships straight. European history is tangled with numerous cross-connections between families and generations. One of the reasons I have included the names and dates of parents and children is to assist the reader, not in memorizing this information, but to be able to cross check associations if needed. (I trust the reader to avoid information overload!) Mary's parents (married in 1551) were Sir Henry Sidney (1529-1586) and Mary Dudley Sidney (1530/35-1586).[46] Mary Dudley's father, John Dudley, Duke of Northumberland had plotted to put Jane Grey on the throne. When all was said and done, Henry Sidney was able to get back on the good side of Queen Mary. Mary Sidney Herbert had numerous brothers and sisters although most died young. Mary was the fifth child: Philip (1554-1586), Margaret (1556-1558), Elizabeth (1557) dies an infant, Ambrosia (1560-1575), Mary (1561), Robert (1563-1626), and Thomas (1569-1595).

Despite Henry Sidney's service to Queen Elizabeth I as Justice and Governor of the Kingdom of Ireland and Lord President of the Marches of Wales, he was not compensated very well and the family struggled with finances.[47] Similar to Margaret Clifford, Mary's mother, Mary Dudley was close to Elizabeth I. Mary Dudley actually contracted smallpox while nursing Elizabeth I. Mary Dudley's brother (and Mary Sidney's uncle), Robert Dudley, was a favorite of the Queen. It is believed that there would probably have been a marriage if not for the suspicious circumstances in which his wife died.[48]

Mary Sidney Herbert was given a robust education: Scripture, the classics, rhetoric, French, Italian, Latin, some Greek and Hebrew but evidence sketchy on the latter. Likewise, she studied music, needlework, and "household medicine."[49] She joined the Royal household of Elizabeth I in 1575 and married, a few years later in 1577, Henry, Earl of Pembroke.

Due to the family's financial difficulties, Uncle Robert arranged Mary's marriage to Henry Herbert with a dowry of £3000. Mary's children were William Herbert (1580), Katherine Herbert (1581), Anne Herbert (1583), and Philip Herbert (1584). Katherine and Anne died early, Katherine when Philip was born (she was about 3 years-old) and Anne in her twenties.

Keeping in mind the kinds of stories that could bloom around women who did not follow a typical path, Mary is described by John Aubrey (a contemporary and member of the Royal Society) as beautiful and witty. He adds that she was also said to be "salacious"[50] and the story was told that she enjoyed watching her horses mate as a prelude to having intercourse herself. This sounds like the kind of gossip to which Margaret Cavendish was subjected, a matter that will be discussed shortly. Women might be educated but if they strayed from certain social mores, they were fodder for unkind talk. In fact, Mary and her brother Philip were so close that it was rumored that they had a sexual relationship and that her first baby was fathered by him. However, even the gossip collector Aubrey was more inclined to disbelieve that story, chalking it up to gossip of "old Gentlemen."[51]

Mary's brother, Philip, gained recognition writing mainly sonnets and poetry but died in 1586 in the Netherlands in battle against Spain. Mary was devastated. She was not allowed to include her thoughts about him in the elegies published by Oxford, Cambridge, and Leiden, nor was she allowed to take part in the funeral due to her sex, perhaps because they were male only universities? However, it is also reported that Mary went on to write and publish as well as serve as a patron to several family members and acquaintances. Philip had taught Mary alchemy and after his death, she carried on his work and patronage of a number of well-knowns such as Edmund Spenser, John Donne, and Ben Johnson. Mary Sidney Herbert's focus seems to have been mainly literature, not science, but she was certainly part of an alchemical circle. Mary died in 1621 from small pox and is buried in Salisbury Cathedral.

Mary Sidney Herbert's Alchemical Associations

Mary's older brother, Philip (1554-1586), was dear to Mary and vice versa. Philip was a favorite nephew of Robert Dudley and in 1577, the uncle, nephew and Edward Dyer traveled to Mortlake to confer with John Dee on their work in natural philosophy. Philip is believed to have studied alchemy which is consistent with his Dee relationship.[52] John Dee wrote the *Monas Hieroglyphica* and truly believed in transmutation—learning about alchemy was a way to understand God.

Upon the death of her first husband, Mary married Matthew Lister, a Knight who belonged to the College of Physicians in London. Mary and Lister built Houghton Lodge in Bedfordshire near Ampthill. Aubrey describes the house as one that was built to replicate the house of Basilius, a character in *Arcadia*, a story begun by her brother that was revised much by Mary.

Mary became known for her studies in chemistry being referred to by Aubrey as "a great Chymist, and spent yearly a great deale in that study."[53] Adrian Gilbert (half-brother to Sir Walter Raleigh, also a chemist and colleague of John Dee) served as her "laborator."[54] This does not seem inconsistent when one begins to see how John Dee was very close to the same people who were close to Elizabeth I, such as the Dudleys, Cliffords, and Sidneys. In fact, it is believed

John Dee tutored Mary's mother and uncle (Mary and Robert Dudley) when her grandfather concluded that they needed to be well schooled in the sciences.[55]

John Dee was part of a group of aristocrats during Elizabeth I's rule, referred to as the Sidney Circle. The Circle included Philip Sidney (Mary's brother), Edmund Spenser, Edward Dyer (alchemical experimentation), Daniel Rogers, John Dee (famed alchemist), Sir Frances Walsingham, and Thomas Moffett who served as the chief physician to the Pembroke family. John Dee (1527-1608) was a mathematician and an astrologer (Astrologer Royal) for Queen Elizabeth I and as well, did work in navigation important to the Queen.[56] It is believed that Dee taught chemistry to Philip. One cannot underestimate the influence Dee had on the men and women practicing alchemy, both during his time and after.

Dee's diary provides insight into the thinking of this gentleman whose ideas that merge science and the occult were thought of with great credibility. He records in his diary hearing knocking and other sounds in his bedroom at night as well as interesting but disturbing dreams.[57] The following is a story that illustrates the mingling of science and theology prevalent in Elizabethan thinking, as well as Dee's work. On Aug. 2, 1590 Dee references one of the children's nurse, Ann(e) Frank (he spells Anne both ways). He mentions her 'great affliction of mynde."[58] On Aug. 22 Dee writes, "Ann my nurse had long byn tempted by a wycked spirit: but this day it was evident how she was possessed of him. God is, hath byn, and shall be her protector and deliverer! Amen."[59] A few days later Dee describes Ann as being better. However, on Aug. 26 Dee records that he anointed Ann with a holy oil "in the name of Jesus"[60] Ann needed further treatment and Dee writes on Aug. 30:

> In the morning she required to be anoynted, and I did very devowtly prepare myself, and pray for vertue and powr and Christ his blessing of the oyle to the expulsion of the wycked; and then twyse [twice] anoynted, the wycked one did resest a while.[61]

On September 8, Ann attempted to drown herself in a well but Dee was able to save her in time. One can only imagine the depth of what Ann was experiencing in these times when depression and melancholy were thought to be brought on by supernatural or evil forces. At any rate, on September 29, Dee records her tragic but successful suicide.

> Nurse Anne Frank most miserably did cut her owne throte, afternone abowt four of the clok, pretending to be in prayer before her keeper, and suddenly and very quickly rising from prayer, and going toward her chamber, as the mayden [maiden] her keper thowght, but indede straight way down the stayrs [stairs] into the hall of the other howse, behind the doore, did that horrible act; and the mayden who wayted on her at the stayr-fote [stair-foot] followed her, and missed to fynde her in three or fowr places, tyll [until] at length she hard [heard] he rattle in her owne blud.[62]

Dee also records the visit of the Countess of Cumberland, Margaret Clifford, on Dec. 3, 1593 and indicates that it was not her first visit.

Dee spent time in Prague where King Rudolf II, who was absorbed by the occult, Hermetic arts, and natural philosophy, supported numerous practitioners. There is much to be said about John Dee but for the purpose of this book, I will focus on his beliefs since they were studied by Philip Sidney and very likely, Mary.

Dee believed that it was possible to communicate with angels if he could only learn how. He craved esoteric knowledge which was thought to enlighten one about God but was not to be found in books. He cites the use of a "Shew-stone"[63] that is used to make the link for communication. Dee employed *skryers* such as Edward Kelley, who gained some notoriety as a probable fake. Kelley would channel the angelic information to Dee who recorded it in detail. The angels, Annael and Uriel, told Dee/Kelley how to construct the sigils, table, ring and lamen (pendant or breastplate) needed for the work. This went on for a long time. By 1582, Dee's writing implemented what was called the Enochian language. It is a remarkable code that goes on for pages and protects the information from unworthy readers.

John Dee wrote profusely and his work, *Monas Hieroglyphica* published in 1564, became a must-have book for alchemists. It is said he wrote it in seven days. It contains a mixture of mathematics, Kabbalah, and alchemy explained via theorems. The most famous figure is the actual Monad which is a combination of the four major elements as they were known at that time. (See figure 6.2).

    ⌐ Luna (moon)

    ⌐ Sol (sun)

    ⌐ Elementa (components)

    ⌐ Ignis (fire)

Figure 6.2: Theorem 10 Shows the Monas Symbol Explained Where He Shows Combined and Broken into the Four Elements.[64]

The *Monas* contains quite a few charts and diagrams that explain Dee's mystical interpretations about the nature of the universe. The existence of the elements is tied to God's creation as well as the life, death, and resurrection of Christ. He also includes a few diagrams that explain addition and multiplication via his understanding of Pythagoras.

Dee had a personal collection of books and manuscripts, over 2500, that included mathematics, alchemy, magic, philosophy, the Kabbalah, geography, cryptography, and literature. Despite the patronage of Elizabeth I, Dee had to be very careful in the end.

The last six years of his life were spent under James I, an unpleasant situation for anybody remotely suspected of witchcraft, this monarch having been responsible for the most savage persecution of witches and magicians ever to occur in England.[65]

Therefore, Mary Sidney Herbert may not have produced hard evidence that she practiced alchemy but her reputation was that of being an intellectual, beautiful, woman who did know *chymstry*. Her circle of family and friends were at the forefront of alchemical study in their day.

Anne Conway: Prima Materia and Consciousness

Sarah Hutton argues that two seventeenth century women, Anne Conway and Margaret Cavendish, have been ignored by traditional historians who have not recognized their contributions to the Scientific Revolution. Anne Conway (c. 1600-1677) is described by Hutton as one of the "most highly educated women of her generation, in so far as she was the lucky recipient of the equivalent of a university education."[66] Anne engaged in a life-long correspondence with Henry More, her brother's tutor at Cambridge. More is also well known as one of the famous Cambridge Platonists, a group who eventually included Newton. Anne Conway was only published posthumously, a significant element in the story of the *soror mystica,* as it is another example of the invisibility of women in the history of science. Anne Conway studied Descartes and Plato and whereas she, like many of her sisters, acknowledged much of Descartes' work, she could not accept a dualistic nature of body and spirit.

A few biographical notes will help the reader understand the times in which Anne lived. Anne Finch Conway was born in December 1631 in London.[67] She suffered from migraine headaches throughout her lifetime that were believed to have been brought on by an illness she contracted at the age of 12 years. The family thought that as well, her headaches were due to Anne studying too much. It is interesting how often a perception of too much study was pathologized for both women and men in this century. Although Anne's education was not a traditional one in the formal sense, her brother, John Finch, kept her supplied with books during the course of his own studies at Cambridge such as the work of Descartes. John's teacher, Henry More, began a 30-year correspondence with Anne at her request. He translated Descartes (his concentration of study) for Anne but both teacher and student eventually rejected parts of Cartesian philosophy. Henry More also studied the Kabbalah. However, it appears that More's study of the Kabbalah was with the intent to convert Jews to Christianity, not on an interest in esoterica.[68] Waite characterizes More as an evangelist which is consistent with the Millenarian movement that will be discussed in the next chapter and in which More was involved.

Anne married Edward Viscount Killultagh, first Earl of Conway when she was 19-years-old. During her marriage, Anne continued her studies in math, geometry, and astronomy. In 1660, Anne's only son died from smallpox; she also contracted it and survived, but was in pain for the rest of her life. Her infirmity and chronic pain led Anne to devote much of her work to finding a cure for her condition. At one time, Samuel Hartlib's son-in-law, Frederick Clodius, performed a procedure on Anne in which she almost died from mercury poisoning. She also nearly succumbed to the temptation to try a procedure called trepanning which consisted of drilling a hole through the skull to relieve pressure. Fortunately, she did not undergo the operation.

The Conways moved to Ragley Hall in Warwickshire, where Henry More eventually also took up permanent residence. Anne Conway had a laboratory at her home in Ragley in which she experimented in natural philosophy. She was well versed in alchemy as were Henry More and another close friend, Francis Mercury van Helmont, son of the famous alchemist and physician, John Baptiste van Helmont. Anne's brother, John, suggested that Francis might be an invaluable source as an aide for her migraines (circa 1654). Francis Mercury van Helmont was both an alchemist and editor of a publication titled *Kabbala Denudata*. Van Helmont traveled to England in 1670 with letters of introduction written by Elizabeth of Bohemia to Henry More. Subsequently, Van Helmont joined Anne's group of friends, taking up residence at Ragley Hall and setting up a lab with More. In addition to alchemy, both men played significant roles in Anne's subsequent study of Kabbalah and the development of her theory of monads.

Hutton believes Katherine Boyle Jones and Anne Conway may have known each other, although there is no evidence in the extant letters. She suggests that Katherine's involvement with the Hartlib Circle, her knowledge of Hebrew, and her vast intellect were commonalities held by these two gifted women who were about 14 years apart in age. As mentioned earlier, it is difficult to really know Katherine Boyle as her letters have either gone missing or did not survive. Conclusions drawn about her come from scholars' analysis of her political activities as well as her known associations with Hartlib members such as John Milton and her brother, Robert Boyle. Conway also knew Robert Boyle so it is logical to speculate that the two women were acquainted. They had other mutual acquaintances, so at least, must have known of each other's existence. Hutton writes, "On at least one occasion Lady Conway instructed More to send paper to her via Lady Ranelagh's house in Pall Mall." [69] They also shared an interest in alchemy.

Theory of Monads

Rene Descartes' theories have been the source of controversy and blamed for leading to the separation of matter and spirit. Put simplistically, Descartes believed in a universe that consisted of particles that behaved like parts of a machine, with humans acting as the carriers of a soul that was separate from matter. Newton added a more formal mathematical framework for how molecules behaved. However, it was Conway (and later, William Leibniz) who conceived of molecules as monads that had a life force of their own, not simply parts of a great mechanism. She stated that all creation was linked in a "Great Chain of Being."[70] The chain manifested from a *prima materia*, again, a sort of origin of existence, and then grew in complexity to result in humanity. Think of a great soup of elementary particles that self-organize into organisms of greater and greater complexity. Some organisms halt at being a tree or a toad. Others keep organizing into human beings. However, if all we know that exists in our physical world is reduced to its smallest self, we see that we are all made of the same *stuff* and furthermore, are interconnected. The reader may notice that again, we hear elements of alchemy and Kabbalah. Recall that the Kabbalah taught that all matter emerged from a type of first matter whose origin is God.

A belief in transmutation would be a logical outcome of the chain's continuity. Matter could be transmuted from simple to complex, from *prima materia* to humanity, from lead to gold. The monad argument continued between proponents of a matter-based universe and between those of a spirit-filled universe. For example Emilie du Châtelet, lover to Voltaire, wrote about *forces vives* [living forces], in her superior French translation of Newton's *Principia* in the 1730s.

Anne's work in alchemy and her subsequent writing are excellent examples of the overlap of scientific thinking and theological beliefs that are found often in alchemical literature. According to Margaret Alic, Kabbalism provided Anne with another source for her thinking about the unity of nature with its, as well as alchemy's, focus on the union of the male and female principles, as well as the argument for their equality. Anne concluded that considering people were made in the likeness of God, matter not only behaved according to Descartes' mechanistic notion of the universe but in a sense, had a kind of consciousness. She wrote in her treatise, *The Principles of the Most Ancient and Modern Philosophy*:

> For truly in nature there are many operations that are far more than merely me-chanical. Nature is not simply an organic body like a clock, which has no vital principle of motion in it; but is a living body which has life and perception, which are more exalted than a mere mechanism or a mechanical motion.[71]

Anne's philosophy concerning the unity of spirit and matter is referred to as "vitalistic natural philosophy."[72] She writes of a "Middle Nature" in which Christ is a mediator between God and that which he created. Hutton summarizes from Conway's *Principles*:

> By sharing both the attributes of God (such as immutability) and attributes of created things (mutability), Middle Nature is the mediator between being and becoming, between the uncreated and the created, between the eternal and im-mutable deity and the temporal and changeable world.[73]

Christ seems to be both Son of God and the first created being.

Around 1677, Anne converted to Quakerism to the dismay of her family. It was Francis van Helmont who introduced her to the Society of Friends. Quaker belief is consistent with the Kabbalah as both belief systems argue that each person carries a part of God or the Inner Light. The misunderstanding occurs when one confused carrying a part of God as something physical rather than spiritual. Jacob Böehme also taught this thinking and it was not unusual for those who studied natural philosophy to spend time examining Quaker belief. Anne's intellectual circle believed in the goodness of God and they were convinced that the ancient practices were ideals to follow, known as the *priscia sapientiae* or ancient knowledge. One of Anne's significant teachers in Quaker tradition was George Keith who made use of alchemical terms in a Jungian-like association to suffering. He describes the process of obtaining perfections of the spirit, "'As by calcination, melting in the fire, heating it, pounding and pressing it, bruising and squeezing, boyling and many such things, wel known to Chymists and Physicians.'"[74]

The Quaker tradition holds similar beliefs to Baruch Spinoza's conviction that the inner Divine light was to be found in all people (he spent some time studying with the Quakers, as well). That there is something of God in every man bothered many who saw it as a literal idea that could only be blasphemous. It seemed they were saying that Christ was the seed, literally, in a person and that sounded too physically intimate. Speaking from that inner light was certainly a threat to the traditional Church structure as the Quaker thinking elevated the common person to one who was worthy of studying and interpreting Scripture. That was a threat to the power structure or Church, who would not be able to control the results of people coming to their own conclusions about Scripture. The charge of heresy was always in the background. Recall that in 1545-1563 the Catholic Church based Council of Trent wrote decrees including one that stated, "'no one, relying on his own judgment and distorting the Sacred Scriptures according to his own conceptions, shall dare to interpret them.'"[75] This is mindful of Newton's interests that ranged beyond that of science to the theological. Newton subscribed to a belief system called Arian philosophy in which Christ and God are not *One* as is traditionally taught regarding the Christian Trinity. Rather, Christ is the first creature created by God.[76] Newton kept these beliefs to himself as most Christians did not embrace them.

Anne, too, fell prey to the need to accept the union of spirit and matter but to dismiss that this could encompass God the Creator. One of the puzzling things to consider regarding Anne Conway and many of the other seventeenth century women of alchemy concerns their belief in a union of spirit and matter that only goes as far as that which was created by God; yet, did not include God. Anne studied the Kabbalah and accepted much of the idea that reality as we know it, manifests from a great void; however, it appears that her concept of God was of someone separate from the void and in control of it. She and Henry More were quite opposed to Spinoza's philosophy of God which seemed to be pantheistic although he argued vociferously that he was misunderstood.[77] It is difficult for my Jungian influenced mind to understand how one can embrace these seemingly logical fallacies amidst an alchemical conception of a *coniunctio* reality. If one believes that everything emerges from a first matter, is God of that matter or somehow is God in control of it? These are questions that have plagued philosophers, theologians, and scientists throughout history. Even Newton's God was in charge of the universe and separate from it. Perhaps the same inability to accept Spinoza's philosophy of God being omnipresent was at work here. Whereas I believe Spinoza's sense of God is much closer to the meaning of an alchemical universe, for Anne to adopt his theories might have meant taking a step in a direction in which she was not prepared to go. The union of spirit and matter was important but God was something else altogether. Hutton reminds us that Anne Conway was a woman of her times. She studied the new scientific philosophy but in the context of a paradigm that is closer to medieval thinking in a manner that has been argued regarding Newton's limitations. There was a great deal of work that needed to be accomplished in the psyche of humanity as it moved from antiquity to modernity.

Anne Conway died in 1679. Van Helmont returned to Europe with her papers and a notebook in which she had spelled out her monad philosophy. He published it first in Latin in the year 1690 and later in English under the title: *The Principles of the Most Ancient and Modern Philosophy, Concerning God, Christ, and the Creature; that is, concerning Spirit, and Matter in General.* Although Francis van Helmont edited this work by Anne, he took credit for writing it. Alic notes:

> Anne Conway was thus robbed of the credit she deserved because of her sex, just as so many other women have had their work attributed to male scientists, been demoted posthumously to the position of assistant or 'help-mate', or had their very existence denied by historians.[78]

Alic argues that this was not unusual for women. One only needs to consider the way Watson and Crick failed to credit the work of Rosalind Franklin that made their discoveries about DNA possible. She asserts that many male scientists did give credit to the women who were instrumental in their field yet were subsequently dismissed by historians who refused to believe that these women were true scholars. Historians often wrote more about these women's social lives, especially if they lived outside a traditional value structure. Women also published their work under male pseudonyms in order to have it received with credibility. They often lacked the confidence in their work to attempt to present it formally in the scientific community. For example, it is rare to hear Katherine Boyle Jones' name mentioned as anyone other than someone who provided her brother a lovely place to live.

Margaret Cavendish, Duchess of Newcastle: A Genuine Borderlander

The frontispiece that adjoins the title page for *The Philosophical and Physical Opinions, written by her Excellency, the Lady Marchionesse of Newcastle* depicts Margaret Cavendish as an attractive young woman, head cocked a little to the left, wearing a small coronet that appears to be about to topple off her head. A quill and inkbottle sit on a table at her side, next to a blank sheet of paper. Margaret's gaze is in the direction of her precious library. At least we can draw that conclusion as there is an inscription on the picture that reads:

> Studious She is and all Alone
> Most visitants, when She has none,
> Her Library on which She looks
> It is her Head her Thoughts her Books.
> Scorninge dead Ashes without fire
> For her owne Flames doe her Inspire.[79]

Thus, we meet the Marchioness of Newcastle, Lady Margaret Cavendish, who was the first woman to be allowed to visit the Royal Society on May 30, 1667. Eventually she was given the unkindly sobriquet "Mad Madge,"[80] apparently due to her seeming eccentricities, a topic to which I will return shortly. However, Margaret Cavendish cannot be dismissed so easily. As we become familiar with her writing on all manner of scientific topics, it is clear that this woman was curious, enormously intelligent, and liable to make pointed observations about her contemporaries using humor that they did not often appreciate. Her practice of alchemy is more speculative but she is well deserving to be included in this discussion of women alchemists.

Margaret's Early Life

Margaret Cavendish neé Lucas was born in 1623. Her parents were Thomas Lucas and Elizabeth Leighton Lucas, who had caused a scandal by bearing their first son out of wedlock. Thomas died when Margaret was only three-years-old. Margaret describes a happy childhood in her autobiography.[81] Her father had left the family in respectable financial shape and although they were not overly wealthy, they had enough money to live a comfortable lifestyle. Elizabeth Lucas raised the children using reason rather than physical punishment as a disciplinary method. Perhaps due to Elizabeth's earlier difficulties with societal mores (having a child out of wedlock), she kept the family somewhat isolated from society and was a stickler for decorum. She made it clear that there was a distinct boundary between the children and the servants in order to ally her fears that the children might learn uncouth behavior from them. Yet, the children were expected to treat all people with respect. Margaret describes her mother's views on education as somewhat traditional but with her own bias, even obsession, toward cultivating civility. "For my mother cared not so much for our dancing and fiddling, singing and prating of several languages, as that we should be bred virtuously, modestly, civilly, honourably, and on honest principles." [82] Although Elizabeth may have appeared the rebel by becoming pregnant before marriage, either the rebellion was a passing thing or perhaps, the pressure of being a widow with limited means added a practical need to be certain her children fit successfully into society.

Margaret describes herself as someone who grew up thinking about everything all the time. She mentions that she would take walks of two or three hours to think through some idea. However, she also could focus her attention on her true friends and listen to their problems. It was strangers who bewildered her. She observed much and found many folks lacking in civility. She liked toys and clothes but not overly so, something that seems to have changed in her adult life. Margaret had an early education that was typical for a girl of her social class: reading, writing, singing, dancing, and music. She was unable to learn a foreign language finding memorizing vocabulary impossible. Margaret loved to read and when she did not understand a concept, she would ask her brother Lucas to explain it if he could. Being quite shy, she found an outlet in writing that could be described as obsessive. [83] Despite her shyness, Margaret demonstrated a need to communicate her ideas with the world early on and writing allowed her to do so without having to speak to people.

Making an effort to fit into society, Margaret requested that she be allowed to serve as a maid of honor to Queen Henrietta Maria in 1643. She had to persuade her mother to be allowed to go but her brothers and sisters were much more reluctant, being aware that she was both inexperienced and shy. How would she be able to succeed in a position that required tact, people skills, and some worldly wisdom? Margaret is candid about her own nature writing:

> I was so bashful when I was out of my mother's, brothers', and sisters' sight, whose presence used to give me confidence . . . . that when I was gone from them, I was like one that had no foundation to stand, or guide to direct me, which made me afraid, lest I should wander with ignorance out of the ways of honour, so that I knew not how to behave myself. Besides, I had heard that the world was apt to lay aspersions even on the innocent, for which I durst neither look up with my eyes, nor speak, nor be any way sociable, insomuch as I was thought a natural fool. Indeed I had not much wit, yet I was not an idiot. [84]

A shy personality did not bode well for dealing with the machinations of life at Court. Margaret wrote about trying to overcome her bashfulness but it was to no avail. She states that she was not prone to blushing but her anxiety "contracts my spirits to a chill paleness." [85] Margaret naively believed that people she considered to be wise or learned would not judge her harshly, thus she often tried to employ a kind of mental game, imagining that someone she met, whom she believed to be a fool, was actually wise, which might diffuse that person's intimidating authority. Margaret regretted her decision to live at court. The experience was overwhelming to someone with such a reserved nature and she asked to be allowed to come home. Her mother thought this unbecoming, saying that it would look inappropriate to return home so soon. Thus, Margaret accompanied the Queen to France when she went into exile after the beheading of her husband, Charles I. It was in Paris that Margaret met the man who was to become her husband, William Cavendish.

We learn much about Margaret's marriage to William Cavendish in the biography she wrote about him. Margaret describes William as her only love. William was 30 years older than Margaret and a widower. She explains how she met William for the first time when he came to visit the exiled Queen Henrietta Maria in Paris in April 1645. Eileen O'Neill portrays their romance as one carried out mainly through an exchange of letters; however, it did move forward and they married. William Cavendish is described as an "amateur scholar"[??] who studied philosophy, alchemy, poetry, and astronomy. Margaret wrote that William liked her just as she was. Finally, here was a person who admired her intellect as well as her beauty. Despite her dread of men, she fell in love with him. Perhaps his being older gave her a comfort that younger men could not.[??]

The truth of Margaret and William's relationship depends on the observer, or of the historian. According to Margaret's autobiography, it sounds as if the marriage was not one of passion but that of two good friends. After reading her remarks disparaging romance literature, it would seem she had an aversion to romantic passion. According to other sources, William had been known to be a womanizer who became impassioned with Margaret's comely physique, hoped to have a harmless affair with her, and was given a resounding answer of no.[88] These two characterizations illustrate the difficulty in conducting sound historical research. Reading primary sources provides the voice of the historical person with his or her own conscious, as well as unconscious complexes. The same can be said for historians who filter information through their own complexes and projections. The best we can probably do is at least present differing accounts as well and search for corroboration. Needless to say, Margaret admired William greatly for being willing to go into exile with the Royal family out of loyalty to King Charles, an act that resulted in the loss of much of his wealth. Margaret's own family was also ruined by maintaining their loyalty to the Crown. Their estates were seized and the family wealth appropriated. The couple would eventually return to England with the royal court in 1660 when Charles II was asked by Parliament to retake his place on the throne.

William had hoped that by marrying Margaret, he would have the opportunity to father more sons; however, she was unable to have children. Despite that disappointment, William appears to have remained a loving husband. He had lost almost everything in the war. While living in exile, he was living on credit and there came a time when the couple became so desperate, Margaret relates a story about having to pawn some toys she had given her maid. Yet,

Margaret and William seemed to be able to create a loving family life that included his brother, Charles Cavendish.

Margaret loved her brother-in-law, deeply, describing him with much praise. Sir Charles, a skilled mathematician, was a fellow outsider as he suffered from dwarfism or achondroplasia. Cavendish's curved back is mentioned in *Aubrey's Brief Lives*. "'He was a little weake crooked man, and nature having not adapted him for the court nor campe, he betooke himselfe to the study of the mathematiques, wherein he became a great master'"[89] Charles received his knighthood in 1619. He was unmarried and buried in the family graveyard at Bolsover. Both William and Charles supported Margaret in her own studies which must have been highly affirming for her.

Margaret sounds like an archetypal introvert. She did not really like going out socially but had a passion for thinking and writing. She described herself as melancholy rather than merry. She was happy to be near William, her source of support, and to pursue her intellectual interests. Margaret, William and Charles formed an intellectual circle known as the "Newcastle Circle"[90] that included thinkers such as Thomas Hobbes and Kenelm Digby, as well as the occasional participation of Rene Descartes, Pierre Gassendi, and Marin Mersenne. Margaret is described as a "diamond" who was curious about everything. For example, it is said that she pondered "'whether it be possible to make men and other Animal creatures that naturally have no wings, flie as birds do?'"[91] She also wondered if vision originates in the eye, in the brain, or in both.

Margaret was a prolific writer who did not always take the time to revise or edit and was often mocked by her contemporaries. Reading the preface of *The Life Of William Cavendish Duke of Newcastle To Which Is Added The True Relation of My Birth, Breeding, And Life*[92] is enlightening and reveals Margaret's frustration with traditional academics who dismissed or doubted her ideas. She also exhibits humility and humor. Margaret states that she is slow to anger, not typically jealous, fair, kind, and good-natured. She could be self-righteous but also very forgiving. She worries over much about losing the people she loves. She accepts that she is a timid person but also has no doubt that she would risk her life for her loved ones. However, adventure is not for her. She does seek to be remembered after she dies, I believe, as a scholar. Margaret concludes her preface rather poignantly by apologizing to those who would criticize her for writing a biography, imagining they would think her vain and over-confident to believe people would want to read about her life. Yet, she states that she wrote the book for herself, so that "after-ages"[93] would know that she was the daughter of Lucas of St. Johns and as well, the second wife of William Cavendish, not to be mistaken for his first wife or for any succeeding wife. She wants to have a voice. This is who she is.

Margaret: Not Really Mad at All

Margaret Cavendish had a rather polarizing effect on people.[94] Margaret loved discussing philosophy, science, and literature. She also craved the acceptance of men whom she deemed to be learned and envied their freedom to explore the natural world unfettered by social mores. Thus, many men enjoyed her company and many women resented it. Margaret was more inclined to speak about the studies that interested her rather than about local gossip. She tended to chatter when she became nervous and was so concerned that her guests would be comfortable, overdid the amenities. At one time she heard that she was being criticized by guests for avoiding talk about her writing. Upon trying to rectify this, she was then criticized for talking about her work too much. Mary Evelyn (wife of the famous diarist, John Evelyn) was especially harsh, writing in her diary about one of their visits:

> My part was not yet to speak, but admire; especially hearing her [Margaret] go on magnifying her own generous actions, stately buildings, noble fortune, her lord's prodigious losses . . . what did she not mention to his or her own advantage? . . . Never did I see a woman so full of herself, so amazingly vain and ambitious.[95]

A few notes about Mary Evelyn will help place the criticism of Margaret Cavendish in context and illustrate why she was doomed to come up short in society. Mary Evelyn was educated but clearly convinced of a woman's place in society. She disapproved of Margaret, not only because she discussed science and politics but also because Margaret wrote about issues that Mary Evelyn believed were the purview of men. Reverend Dr. Ralph Bohun, an Oxford scholar, wrote a glowing portrait of Mary Evelyn's character. An excerpt from a letter written by Mary Evelyn to Bohun sheds light on the culture in which Margaret was trying to fit.

> January 4, 1672: Women were not borne to read authors, and censure the learned, to compare lives and judge of virtues, to give rules of morality, and sacrifice to the Muses. We are willing to acknowledge all time borrowed from family duties is misspent; the care of children's education, observing a husband's comands, assisting the sick, relieving the poore, and being servicable to our friends, are of sufficient weight to employ the most improved capacities amongst us. If sometimes it happens by accident that one of a thousand aspires a little higher, her fate commonly exposes her to wonder, but adds little to esteeme. The distaff will defend our quarrells as well as the sword, and the needle is as instructive as the penne.[96]

Mary Evelyn continues to write that although she enjoys reading about many subjects, notwithstanding that she is surrounded by science with her husband a member of the Royal Society; she is content to leave those fields to men.

Keeping in mind that Mary Evelyn once called Margaret a "chimera—a grotesque, terrifying, fire-breathing hybrid monster of ancient myth"[97] it is easy to imagine that Margaret was the person Mary had in mind when penning her thoughts on proper womanhood stating, "I hope, as she [Margaret Cavendish] is an original, she may never have a copy."[98] Margaret Cavendish represented so much of what Mary Evelyn feared and perhaps, envied.

To be fair to her detractors, Margaret did seem to be a cause for wonder as some of her outfits were quite bold and fantastic. She was said to have worn a dress on a visit to Charles II that was much talked about, but that may have been due as much to its extravagance, as to the faux pas she made by having a female attendant carry her train, the prerogative of only the most high ranking woman. Margaret recalls:

> I took great delight in attiring, fine dressing, and fashions, especially such fashions as I did invent myself, not taking that pleasure in such fashions as was invented by others. Also I did dislike any should follow my fashions, for I always took delight in a singularity, even in accoutrements of habits. But whatsoever I was addicted to, either in fashion of clothes, contemplation of thoughts, actions of life, they were lawful, honest, honourable, and modest.[99]

Another famous diarist, Samuel Pepys seemed to be both fascinated and repelled by Margaret. His diary entries describe her clothes in a way that brings to mind modern fashion critics.

> April 26, 1667: Met my Lady Newcastle going with her coaches an footmen all in velvet; herself (whom I never saw before), as I have heard her often described (for all the town talk is now-a-days of her extravagancies), with her velvet-cap, her hair about her ears; many black patches, because of pimples about her mouth; naked-necked, without any thing about it, and a black just-au-corps. She seemed to me a very comely woman; but I hope to see more of her on May-day.[100]

The *juste-au-corps* was a knee-length coat worn by men so perhaps, Margaret caused a stir by wearing men's clothing, grounds for criticism of women even through the 1960s. Yet, the reader may have noticed that Pepys looks forward to catching another glimpse of Margaret.

It is clear that Margaret Cavendish appeared to not only elicit criticism but vituperative rhetoric of the kind that causes a depth psychologist to pause and wonder. Her critics did not just dislike her, they loathed her and left no doubt in their writing as well as their public reactions to her presence. Prior to her visit to the Royal Society, John Evelyn wrote in his diary on April 18, 1667:

> I went to make court to the Duke and Duchess of Newcastle, at their house in Clerkenwell, being newly come out of the north. They received me with great

kindness, and I was much pleased with the extraordinary fanciful habit, garb, and discourse of the Duchess.[101]

Evelyn may have wondered and even made a joke about her attire but seemed to also like the Duchess. Yet, on the day of her visit to the Royal Society on May 30, 1667, the first visit to this venerated institution by a woman, John Evelyn seemed to be more consumed with critiquing Margaret's attire and intellect. He remarks:

> To London, to wait on the Duchess of Newcastle (who was a mighty pretender to learning, poetry, and philosophy, and had in both published divers books) to the Royal Society, whither she came in great pomp, and being received by our Lord President at the door of our meeting-room, the mace, etc., carried before him, had several experiments showed to her. I conducted her Grace to her coach, and returned home.[102]

Not to be outdone, Pepys shows a great deal of disdain regarding Margaret's visit as he writes:

> May 30, 1667: Where I find much company, in expectation of the Duchesse of Newcastle, who had desired to be invited to the Society; and was, after much debate, pro and con, it seems many being against it; and we do believe the town will be full of ballads of it.

> Anon comes the Duchesse with her women attending her; among others, the Ferabosco [a friend of Cavendish], of whom so much talk is that her lady would bid her show her face and kill the gallants. She is indeed black, and hath good black little eyes, but otherwise but a very ordinary woman I do think, but they say sings well.

> The Duchesse hath been a good comely woman; but her dress so antick [antique] and her deportment so ordinary, that I do not like her at all, nor did I hear her say any thing that was worth hearing, but that she was full of admiration, all admiration. Several fine experiments were shown her, of colours, loadstones, microscopes and of liquors: among others, of one that did, while she was there, turn a piece of roasted mutton into pure blood, which was very rare.

> After they had shown her many experiments, and she cried still she was full of admiration, she departed, being led out and in by several Lords that were there; among others Lord George Barkeley and Earl of Carlisle, and a very pretty young man, the Duke of Somerset.[103]

Perhaps Margaret had offended some of the Society, as she did not tread lightly in her discourse. She had criticized both Robert Hooke and Robert Boyle, Fellows of the Society, in her book *Observations on Experimental Philosophy*, which lends a certain irony to the experiments they planned for her! Margaret seemed to believe in the power of the human mind to observe and reason whereas creating experiments seems an invalid and useless occupation. Echoing the earlier critics of scientific method she writes, "Neither ought artists, in my opinion, to condemn contemplative philosophy, nay, not to prefer the experimental part before her . . . But our age being more for deluding experiments than rational arguments."[104] She was distrustful of the information gained by using microscopes and telescopes, observations that would have certainly annoyed the Royal Society members except they could dismiss her ideas as rubbish. Whitaker suggests that Margaret was so overwhelmed by the whole experience that she could not say much more than how much she admired their work. However, it seems that the Fellows hoped for more loquaciousness on her part and were disappointed.

MacCarthy makes several different observations regarding Margaret's so-called eccentricities. Discussing the double standard applied to women writers of the time, she argues that some women were accepted by their social circle, especially if they were related to a well-known male writer or had a family member who was part of the higher social strata. Some women were dismissed by calling their morals into question. Others, such as Margaret, were labeled "queer."[105] Margaret was high born but did not fit into society norms. Interestingly, Margaret's critics continued to be fixated on her dress. The Count of Grammount recorded a story in his memoirs that illustrates what people thought of Margaret.

> As I was getting out of my chair, I was stopped by the devil of a phantom in masquerade . . . It is worth while to see her dress; for she must have at least sixty ells [unit of measure, about a cubit] of gauze and silver tissue about her, not to mention a sort of pyramid upon her head, adorned with a hundred thousand baubles.
> 'I bet,' said the king, 'that it is the Duchess of Newcastle.'[106]

Another story about Margaret comes via a Dr. Wilkins, Bishop of Chester who was discussing quite seriously, his plan for journeying to the moon. He told Margaret Cavendish that the journey would take 180 days to which she asked, how he was planning to be able to feed his horses on the way. I wonder if she was joking but he thought she was odd and perhaps a killjoy as he is said to have later stated that he was surprised she would ask such a question as she was one "who had built so many castles in the air, she might lie every night in one of her own."[107] Margaret inhabited an imaginative landscape in her thinking that was just too unfathomable to her peers; it is a place known to many who inhabit what is called, the Borderland.[108] Borderlanders see the world in ways that are different from their peers. They see the negative space first when viewing art. They hear the voices other people fail to acknowledge. They may focus their energy into work such as Newton did or live quietly and anonymously. Borderlanders often believe *they are the crazy one* but time often proves them to have been just a little more observant.

Margaret's Alchemy

Margaret Cavendish was influenced by a variety of philosophers whom she reflected upon before proceeding to draw her own conclusions. Her alchemical
practice is in her thinking. Her science might be inaccurate but she was thinking
about everything—biology, astronomy, motion, optics, mathematics, and chemistry. Margaret was fluent in natural philosophy, commenting on John Baptiste
von Helmont (father of Francis Mercurius), John Dee and numerous other natural philosophers/alchemists. She read widely and it is really no wonder she
rubbed people the wrong way because her criticisms are pretty pointed. She
eventually published 21 works focusing on the subjects of natural philosophy,
poetry, drama, oration, and fiction. Perhaps one reason she was considered ec-
centric was her so-called seeking of fame. O'Neill notes that, "In an era in which
anonymous authorship for women was standard, Cavendish adamantly published under her own name."[109] Yet, her writing and ideas were frequently dismissed by both her contemporaries as well as subsequent historians. The critique
centered upon questions regarding her rationality and her belief in what was
called "vitalistic materialism."[110] Margaret believed not only in self-moving,
self-knowing, matter but that there existed a kind of *prima materia* or pervasive
underlying matter in the universe. Margaret believed that first matter or first
motion was infinite, having no beginning. It sounds very much like Spinoza's
ideas of first matter as well as Anne Conway's monads.

I have not found evidence that Margaret's acquaintance with alchemy was
anything beyond theoretical. She did refer to the Philosopher's Stone in a description of her husband, William. She mentioned that he was methodical and
dedicated to study.

> And though he [William] hath taken as much pains in those arts [horsemanship
> and swordsmanship], both by study and practice, as chymists for the philoso-
> pher's-stone, yet he hath this advantage of them, that he hath found the right
> and the truth thereof and therein, which chymists never found in their art, and I
> believe never will.[111]

Furthermore, it is likely that although Margaret was familiar with alchemy,
she was not convinced that transmutation of elements was possible. In a letter
written circa March 1665 to Lady Anne Conway by her friend and imminent
scholar, Henry More, he adds his voice to the crowd who could not respect her
work. Margaret often sent More copies of her writing, which he read and dismissed. More referred to Margaret Cavendish sarcastically as "this great Philosopher." [112]

> But I am not fallen upon one hand alone, I am spar'd by neither sexe. For I am
> also inform'd that that Marchionesse of Newcastle has in a large book [*Philo-
> sophical Letters*] confuted Mr. Hobbs, Des Cartes, and myself, and (which will

make your Ladiship at least smile at the conceit of it) Van Helmont [the noted alchemist] also to boot.[113]

Pepys made a diary entry in which he, of course, hated Margaret's biography of her husband, writing on March 18, 1668:

> Thence home, and there in favour to my eyes staid at home, reading the ridiculous history of my Lord Newcastle, wrote by his wife; which shows her to be a mad, conceited, ridiculous woman, and he an ass to suffer her to write what she writes to him and of him.[114]

Pepys seems to have detested Margaret beyond reasonable judgment and it is my speculation that she may have spurned him as he was a bit of a lady's man and she was quite comely. Yet, the University of Cambridge and Charles Lamb, specifically, reviewed the book positively. Her publication dates for major works in natural philosophy (plays, poems, orations not included) are:

* *Philosophical Fancies*, London, 1653
* *Philosophical Opinions*, 1655 (a revision of *Fancies*)
* *Natures Pictures Drawn by Fancie's Pencil*, 1656.
* *Philosophical Opinions* (2nd Ed.), 1663.

Perhaps one of the more telling critiques of Margaret's work comes from a much later analysis by Isaac Disraeli quoted here. The italics are mine.

> There are [in the writings of the Duchess] the indisputable evidence of a genius as highborn in the realms of intellect as its possessor was in the ranks of society: a genius strong-winged and swift, fertile and comprehensive, but ruined by deficient culture, by literary dissipation and the absence of two powers without which thoughts are only stray morsels of strength, I mean Concatenation and the *Sense of Proportion*. She thought without system, and set down everything she thought. Her fancy turning round like a kaleidoscope changed its patterns and lines with the most whimsical variety and rapidity. Nevertheless, I believe, had *the mind of this woman been disciplined and exercised by early culture and study*, it would have stood out remarkable among the feminine intellects of our history.[115]

The vituperative reactions to Margaret Cavendish are a little puzzling when one actually reads her work such as *The Philosophical and Physical Opinions*. She exhibits what I interpret as humor, sarcasm, and defensiveness. She rails against readers who think that in order for her to have some written these treatises, she must have spoken at length to, and relied upon, the great thinkers such as Descartes and Hobbes (a member of her intellectual circle). She states that she has read their work but has been more than capable of drawing her own conclusions. The first epistle in the *Opinions* is titled: "An Epistle to justifie the Lady

Newcastle, and Truth against falsehood, laying those false, and malicious aspersions of her, that she was not Author of her Books"[116] She writes:

> First 'tis but your envious Supposition that this Lady must have converst with
> many Scholers of all kindes in learning, when 'tis well known the contrary, that
> she never converst with any profest Schooler in learning, for to learn, neither
> did she *need* [emphasis mine] it, since she had the conversation of her Honor
> able, and most learned Brother [Lucas] from her cradle; and since she was mar
> ried, with my worthy and learned Brother [Sir Charles]; and for my self I have
> lived in the great world a great while, and have thought of what has been
> brought to me by the senses, more then was put into me by learned discourse;
> for I do not love to be led by the nose, by Authority and old Authours, ipse
> dixit [dogmatic thinking] will not serve my turn, were Aristotle made a more
> Philosophical Bible then he is, and all scholers to have a lively faith in him,
> doth not move me to be of their Philosophical churche at all.
>
> And I assure you her [Margaret is speaking about herself] conversation with
> her Brother, and Brother-in-law [Sir Charles Cavendish], were enough without
> a miracle or an impossibility to get the language of the arts, and learned profes
> sions, which are their terms, without taking any degrees in Schooles.[117]

In other words, Margaret's lack of formal degrees does not mean she cannot think and reason well. Margaret defends her knowledge of topics that span a diverse list including theology, transubstantiation (changing of the host into the body of Christ in the Christian Eucharist), astronomy, philosophy, medicine, and physics stating that understanding these subjects is not overly demanding. She refers to the scientific establishment as "the Gown-tribe"[118] and has no difficulty challenging them, a most probable reason for some of their dismissal of her work. For example, she writes that when she discusses "materia prima"[119] and is told by others that it refers to "matter without form"[120] she argues that, "there is no matter without some form, so materia prima are two Latine words that mean nothing."[121] They saw her response as unsophisticated but it has a certain logic to it.

In *Observations on Experimental Philosophy*, Margaret refers or paraphrases her target and then explains what she finds unreliable in their theories. For example, Chapter 32, or "Of the Celestial Parts of the World; and Whether Thy Be Alterable," Margaret tackles an issue addressed by astronomers before her regarding the possibility that change could occur in the heavens. She poses the question, "Whether the celestial parts of the world never alter or change by their corporeal figurative motions, but remain constantly the same, without any change or alteration?"[122] The reader will remember that Aristotle argued that no change occurred in the heavens whereas later thinkers such as Copernicus, Brahe, and Galileo suggested otherwise, not always to an accepting audience.

Margaret determines that change does occur and that one cannot necessarily conclude otherwise from the vantage point of earth. She begins at a micro level, noting that change occurs all the time as seen in plants and animals for instance. However, the earth as a large body, continues to exist seemingly unchanged. "The elements are changed one into another; when as yet the globe of the earth itself, remains the same"[123] Although her science is from the 1660s, she invokes logic when she asks her readers to think of observing the earth from the moon. An observer might see the earth's movement as a planet from such a distance; the everyday back and forth of humans and animals would not be visible. To conclude that there is no change on the earth using one's observations from the moon would be erroneous. She argues that to conclude that there is no change in heavenly bodies would be making the same mistake.

Margaret also writes about her medical knowledge saying that any wife of, for example a farmer or constable, is confronted with those subjects naturally. She argues that if one looks at a sheep that has been cut open, one is going to observe the viscera and understand the anatomy. Moreover, in order to be effective in the healing arts, one is bound to be steeped in the appropriate use of medicine as well as the many diseases and conditions suffered by people. In other words, Margaret makes the case that considering all that women do in their daily lives, it should be no surprise that they have significant knowledge of the natural world without having to refer to well-known scholars such as Galen or Hippocrates. She feels equally capable with the subject of mathematics.

> Now for the great learning of knowing the terms of Geometricians, when this Lady touches upon Triangles, Squares, Circles, Diameters, Circumferences, Centers, lines straight and crooked, etc. I will not dissect these great mysteries, because they are so very common, as the meanest understands all these termes, even to Joyners and Carpenters, therefore surely this Lady is capable of them. [124]

Margaret's theories may seem quaint to modern thinkers but she exhibits a lively, although somewhat practical, intellect. Yet, that assessment may be more an artifact of the times in which she wrote. For example, in chapter 127 of the *Opinions*, she includes a section titled, "The reason of the ebbing and flowing of the sea thus."[125] Margaret agrees with Copernicus and Galileo regarding the earth orbiting the sun; however, she disagrees that the moon causes the tides noting that the tides occur at different times in Scotland and Margell. If the moon caused tides, she reasons they would happen at the same time in these two locations. Margaret also notes that other thinkers believe tides are caused by the sun, or that the salt of the sea gives it heat which causes the tides; the latter does not make sense as it does not explain tidal flow in fresh water.

Margaret's basic thinking was aligned with Hobbes and Descartes in that the universe was believed to be made up of matter that moved and behaved according to its properties.[126] There was no room for metaphysical notions of consciousness in this framework. Yet, Hutton adds, "Unlike the mechanists, Cavendish took a hylozoist position, imputing life, motion and mental powers to body itself . . . . She rejects any idea that Nature might be a manifestation of the divine, when she ridicules [Henry] More's Spirit of Nature."[127] We see here the befuddling of science and theology. Speculating on the nature of the universe required the thinker to remember that God was the architect and humans were mere explorers.

Archetype of the Whore

When a person provokes extremes of either derision or acclaim, depth psychologists like to look at the hooks and projections that lead people to develop such strong feelings about a person they barely know. If the reader has ever met a person whom you immediately loved or disliked, you will understand the phenomenon. Having just met that person, we cannot know him or her but something gets stirred. There is some complex of our own that needs to be considered and the person or subject is often tied to an archetype or basic principle experienced by most of humanity. Louisa Costello wrote about Margaret in 1844 describing women like Margaret Cavendish as, "pedantic and self-sufficient"[128] as if the latter were a wicked trait. Costello continues in a rather dramatic style arguing that Margaret was:

> One of those who, with some talent and no genius, contrive to bring themselves into notice by dint of resolute scribbling—not content to understand and patronize the works of persons of merit—indulge the ambition of imitating them—fondly persuade themselves that they can compete with the best authors of the day."[129]

Margaret Cavendish certainly engendered both harsh criticism and acclaim. Something that emerges for me is what I have already described as passionate, vituperative diatribes against her. The complex fascination and repulsion expressed by Margaret's detractors remind me of a very basic archetype, that of the whore.[130] This archetype is best characterized by the prostitute or loose woman. Women who engender this archetype are often outsiders and dismissed by so-called polite society. Specifically, the descriptions and analyses of Margaret's manners and dress accompanied by the vituperative rhetoric concerning her work, are especially reminiscent of the manner in which the Whore of Babylon is described by John the Apostle in Revelation.

> And I saw a woman sit upon a scarlet-colored beast, full of names of blasphemy, having seven heads and ten horns. And the woman was arrayed in purple and scarlet color, and decked with gold and precious stones and pearls, having a golden cup in her hand full of abominations and filthiness of her fornication: And upon her forehead *was* a name written, MYSTERY, BABYLON THE GREAT, THE MOTHER OF HARLOTS AND ABOMINATIONS OF THE EARTH.[131]

Edward Edinger wrote about the Book of Revelation in his seminal book, *Archetype of the Apocalypse.*[132]He makes an interesting point about the collective psyche. He reminds us that the feminine ideal, embodied in Venus and the older pagan traditions, had to go through its own alchemical transmutation in the evolution of human thinking. He describes the Whore of Babylon as "a degraded version of 'Venus'"[133]who is usually held up as a model of beauty and the ideal woman. His point that will continue to be significant in the next chapter is that we must be careful to avoid concluding that, "certain psychological events in the collective psyche were just errors and could have been different. They could not have been different—they had to be the way they were, given the nature of psychic reality."[134]This does not make Margaret's experience with her peers any less hurtful; however, it does illustrate how these archetypes appear throughout history for a reason.

Margaret is called mad, ridiculous, and conceited by her peers. Her thinking is described as haphazard and her writing is reviled. After her death, a bogus epitaph was circulated in which the concept of whore is described explicitly.

> Here lies wise, chaste, hospitable, humble,
> I had gone on, but Nick [Satan] began to grumble:
> "Write, write," says he, "upon her tomb of marble
> These words, which out I and my friends will warble:
> 'Shame of her sex, Welbeck's illustrious whore,
> The true man's hate and grief, plague of the poor,
> The great atheistical philosophraster,
> That owns no God, no devil, lord nor master;
> Vice's epitome and virtue's foe,
> Here lies her body, but her soul's below [Hell].[135]

A more balanced assessment of Margaret Cavendish would perhaps make note of her eccentricities; yet, even that term is subjective. We may acknowledge that some people found Margaret to be witty; others saw her as vain, silly and old. Usually the truth lies somewhere in between. She appears to have been highly intelligent, somewhat socially awkward but confident enough to publish her work. On the other hand, Margaret was naïve in her giving academia more power than it deserved. She mistakenly believes that university professors will look at her work with open minds and curiosity. She is sadly mistaken. Academia, where learning and great thinking can occur also harbors a culture of jealousy, close mindedness, and cruelty. It also has not changed that much in our modern time.

Margaret's writings in natural philosophy are of course out dated but no more so than some of the reports written by her male peers at the time that one finds in the *Philosophical Transactions.* There appears to be more to the vitriolic comments about Margaret that move us into psychology and interpretation. Therefore we just leave her as a bold soul who was too much of a Borderlander for her contemporaries.

Margaret Cavendish died when she was only 49 years of age. William lived three more years and died when he was about 82. They are buried side by side in Westminster Abbey, a place I visited recently and where visitors are not allowed to take photographs. Distracted by my own projections about Margaret and her struggles as a borderland type personality, I had not even noticed the prohibition to take photos. Thus, a security guard was quick to point out my mistake as I tried to capture a picture of Margaret and William's monument with my digital camera. Seeing that I was overcome with emotion, he kindly directed me to the Abbey librarian from whom I was able to purchase two photos of the monument that were much better than the one I had taken. The inscription reads:

> Here lyes the Loyall Duke of Newcastle and his Dutches his second wife by
> whome he had noe issue, her name was Margarett Lucas youngest sister to the
> Lord Lucas of Colchester a noble familie for all the Brothers were Valiant and
> all the Sisters virtuous This Dutches was a wife wittie & learned Lady which
> her many Bookes do well testifie She was a most Virtuous & a Loveing & care-
> full wife & was with her Lord all the time of his banishment & miseries &
> when he came home never parted from him in his solitary retirements.

They lie under a canopy of stone, heads resting on a pillow. Their arms are intertwined and Margaret is holding a book—of course.

Notes

1. Virginia Woolf quotes a comment about women writing music on page 54 in *A Room of One's Own*, quoted from *A Survey of Contemporary Music*, Cecil Gray, page 246.

2. Margaret Clifford, *Receipts of Lady Margaret Wife of George, 3rd Earl of Cumberland for Elixirs, Tinctures, Electuaries, Cordials, Waters, etc., MS circa 1550 with Her Annotations* (n.p.,1920).

3. H. C. G. Mathew and B. Harrison, *Oxford Dictionary of National Biography*, Volume 12 (Oxford: Oxford University Press, 2004).

4. V. Sackville-West, *The Diary of the Lady Anne Clifford with An Introductory Note by V. Sackville-West* (London: William Heinemann, 1923).

5. Sackville-West, ix-x.

6. Sackville-West, 14.

7. Sackville-West, xxii.

8. Matthew and Harrison, 104.

9. Sackville-West, xxiii.

10. Matthew and Harrison, 79.

11. Sackville-West, xxxii.

12. Sackville-West, 19.

13. Sackville-West, 26.

14. Sackville-West, 28.

15. Sackville-West, 47.

16. Sackville-West, 48.

17. Sackville-West, 48-49.

18. Sackville-West, xl.

19. Sackville-West, xlvii.

20. Sackville-West, 108.

21. Penny Bayer, "Lady Margaret Clifford's Alchemical Receipt Book and the John Dee Circle." *Ambix,* 52 (November 2005): 271-284.

22. Kendal, Cumbrian Record Office, WD Hoth Box 44. *Receipts of Lady Margaret Wife of George, 3r Earl of Cumberland for Elixirs, Tinctures, Electuaries, Cordials, Waters, etc., MS circa (1550) with Her Annotations.*

23. Bayer, 274

24. Bayer, 274.

25. Bayer. 274

26. Bayer, 277.

27. Kendal, Cumbria Record Office. WD Hoth Box 44 4a

28. Clifford, 22.

29. Clifford, 20.

30. Clifford, 27.

31. Theodore Kirkingius, *Basil Valentine: His Triumphant Chariot of Antimony, with Annotations of Theodore Kirkingius, M.D.* (London, Dorman Newman at the Kings Arms in the Poultry, 1678).

32. Rulandus, 32.

33. Clifford, 36.

34. Clifford, 49.

35. Clifford, 53.

36. Clifford, 44.

37. Clifford, 44.

38. Clifford, 108.

39. Baro Urbigerus, *The One Hundred Alchemical Aphorisms of Baro Urbigerus* (Edmonds, WA: The Alchemical Press, 1986/1997). (Originally published in London, 1690)

40. Urbigerus, 10.

41. Lyndy Abraham, *Alchemical Imagery* (Cambridge: Cambridge University Press, 1998).

42. Clifford, 119.

43. Rebecca La Roche, *Medical Authority and Englishwomen's Herbal texts, 1550-1650* (Burlington, VT: Ashgate, 2009).

44. La Roche, 3.

45. Johanna Moody, ed.,, *The Private Life of an Elizabethan Lady: The Diary of Lady Margaret Hoby 1599-1605* (Great Britain: Sutton, 1998): 156.

46. Matthew and Harrison, 708-712.

47. Frances Berkeley Young, *Mary Sidney Countess of Pembroke* (London, David Nutt, 1912).

48. Dick Oliver, ed.,, *Aubrey's Brief Lives* (London: Secker and Warburg, 1949).

49. Matthew and Harrison, 708.

50. Oliver, 138.

51. Oliver,139.

52. Peter J. French, *John Dee: The World of an Elizabethan Magus* (London: Routledge and Kegan Paul, 1972).

53. Oliver, 138-139.

54. French,129.

55 French (1972).

56. Donald C. Laycock, *The Complete Enochian Dictionary: A Dictionary of the Angelic Language as Revealed to Dr. John Dee and Edward Kelley* (London, Askin, 1978).

57. John Dee, *Private Diary of Dr. John Dee and the Catalogue of His Library of Manuscripts*, Edited by James Orchard Halliwell (London: Bowers, Nichols and Son for the Camden Society, M.DCCC.XLII).

58. Dee, 35.

59. Dee, 35.

60. Dee, 35.

61. Dee, 35.

62. Dee, 36.

63. Laycock, 11.

64. John Dee, *Monas Hieroglyphica* (London: Guliel Slivius, 1564), 13.

65. Laycock, 17.

66. Lynette Hunter and Sarah Hutton, eds., *Women, Science and Medicine 1500 – 1700:Mothers and Sisters of the Royal Society* (Gloucestershire: Sutton, 1997), 219.

67. Margaret Alic, *Hypatia's Heritage: a History of Women in Science from Antiquity Though the Nineteenth Century* (Boston: Beacon, 1986).

68. Arthur E. Waite, *The Holy Kabbalah* (New Hyde Park, NY: University Books, 1960). (Work originally published in 1902 as *The Doctrine and Literature of the Kabbalah* with Theosophical Society in London as publisher).

69. Sarah Hutton, *Anne Conway: a Woman Philosopher* (Cambridge: Cambridge University Press,2004), 138-139.

70. Alic, 7.

71. Hutton, 229.

72. Alic, 5.

73. Hutton (2004), 171-172.

74. Hutton, 182.

75. Dava Sobel, *Galileo's Daughter* (New York: Penguin, 2000), 72.

76. White (1997).

77. Robin L. Gordon, "The Murder of Spinoza and Other 17th Century Alchemists: A Contemporary Look at a Long-Ago Mortificatio Tale" (PhD dissertation, Pacifica Graduate Institute, 2004).

78. Alic, 5.

79. *The Philosophical and Physical Opinions, written by her Excellency, the Lady Marchionesse of Newcastle* (Cavendish, 1655)

80. Dorothy Stimson, *Scientists and Amateurs: A History of the Royal Society* (New York: Henry Schuman, 1948), 82.

81.Margaret Cavendish, *The Life Of William Cavendish Duke of Newcastle To Which Is Added The True Relation of My Birth, Breeding, And Life*, C. H. Firth, ed. (London: John C. Nimmo, 1886). (Original work published 1667)

82. Cavendish, 280.

83. Eileen O'Neill (ed), *Margaret Cavendish, Duchess of Newcastle: Observations upon Experimental Philosophy* (Cambridge: Cambridge University Press, 2001), xii.

84. Cavendish, 286-287.

85. Cavendish, 300.

86. O'Neill, xiii.

87. MacCarthy (1944).

88. Mendelson (1987).

89. Quoted in Matthew and Harrison, *Oxford Dictionary of National Biography,* Volume 10, 608-609.

90. O'Neill, xiii.

91. MacCarthy, 82.

92. Cavendish (1667).

93. Cavendish, 318.

94. Katie Whitaker, *Mad Madge: the Extraordinary Life of Margaret Cavendish, Duchess of Newcastle, the First Woman to Live by Her Pen* (New York: Basic Books, 2002).

95. Whitaker, 292-293.

96. quoted in Lord Richard Braybrooke, ed., *Memoirs of Samuel Pepys, Esq.F.R.S.: Secretary to the Admiralty in the Reigns of Charles II. And James II. His Diary from 1659-1669,* 2nd Ed. Volume 3 (London: Henry Colburn, 1828), 434.

97. Whitaker, 293.

98. Whitaker, 293.

99. Cavendish, 312.

100. Braybrooke, 206-207.

101. Austin Dobson, *The Diary of John Evelyn,* Volume II (London: Macmillan, 1906), 269.

102. Dobson, 272.

103. Braybrooke, 229-230.

104. Cavendish, 196.

105. MacCarthy, 21.

106. MacCarthy, 81.

107. Stimson, 83.

108. Jerome S. Bernstein, *Living in the Borderland: The Evolution of Consciousness and the Challenge of Healing Trauma* (London: Routledge, 2005).

109. O'Neill, xi.

110. O'Neill, xxi.

111. Cavendish, 306.

112. Hutton, 237.

113. Marjorie Hope Nicolson, ed. and Sarah Hutton, revised ed., *The Conway Letters: The Correspondence of Anne, Viscountess Conway, Henry More, and Their Friends 1642-1684* (Oxford, Clarendon Press, 1992), 233-234.

114. Firth, viii.

115. MacCarthy, 83.

116. Cavendish, A1.

117. Cavendish, A1-A2.

118. Cavendish, A2.

119. Cavendish, A2.

120. Cavendish, A2.

121. Cavendish, A2.

122. O'Neill, 132.

123. O'Neill, 132.

124. Cavendish, A3.

125. Cavendish, 86.

126. Hutton (2004).

127. Hutton, 115.

128. Louisa Stuart Costello, *Memoirs of Eminent Englishwomen* (London: Richard Bentley, New Burlington Street. Publisher in Ordinary to Her Majesty, 1844), 112.

129. Costello, 211.

130. For more on the archetype of the whore see Edward Edinger *Archetype of the Apocalypse* (Chicago, Open Court, 1999).

131. King James Version (KJV), Revelation 17: 3–5.

132. Edward Edinger, *Archetype of the Apocalypse* (Chicago, Open Court, 1999).

133. Edinger, 133.

134. Edinger, 135-136.

135.Whitaker, 348.

Chapter 7

Alchemy, Kabbalah and the Apocalypse

For truley he that will endevour with successe to finde out the meaning of the Apocalyptique visions, must first of all place the course, and connexion of them one with another according to things done, being thorowly searched out by the foresaid characters and notes, and demonstrated, by intrinsicall arguments as the basis, and foundation of every solid, and true interpretation . . . And the same houre there was a great earth-quake and the tenth part of the citie to (wit that great one) fell, and there were slaine in the earth-quake seven thousand names of men. (Excerpts from *The Clavis Apocalyptica*)[1]

Do you think, O Lover of Chymistry, you understand what you read? You cannot understand, unless either divinely Philip, or humanely Oedipus, appear to you, and clearly teach you the way of preparing this Tincture. [2]

An intriguing story emerged in my research of Katherine Boyle and her circle. In addition to their scientific studies, many of the friends of the Boyles displayed an almost obsessive interest in the Christian Apocalypse. In this chapter, I hope to show how their studies of the physical world were closely tied to their theological searching and furthermore, how this connects to a book about alchemy. Keeping in mind that the study of alchemy had both spiritual and physical aspects, its integration with apocalyptical study illustrates one of the ways in which one branch of study could develop into modern chemistry whereas the other branch remained grounded in spirituality. The separation of these opposite qualities was emerging during the Scientific Revolution.

Alchemy shares a basic tenet with many religious orientations, teaching that humanity is a microcosmic reflection of the macrocosmic cosmos. This macrocosmic reality is not to be confused with the universe as we perceive it with our senses, but encompasses all reality. For example, the reader will recall that many alchemists studied the Kabbalah, the canon of Jewish mysticism. In the spheres of the Kabbalistic Tree of Life, matter is fundamental but is governed by a higher authority. The definition of what constitutes that authority is the subject for another book. Having the basic understanding of how to experiment with alchemical elements of mercury, salt, and sulphur can be compared to attempting to construct a garment with only the knowledge of how to sew a straight seam. The complexity of what goes into sewing a dress with sleeves, darts, zippers and such is vital for success. In a similar manner, a basic understanding of alchemy might be possible; however, it is just a glimpse of a much more complex system. That is how the alchemists saw their work, guided by God, unknowable at a deep level, but worthy of the effort.

I believe it is very difficult for our modern, secular mind to fully appreciate how much our ancestors' thinking about the natural world was intertwined with their theology. Frances Yates puts this well in a statement she made regarding the sixteenth century, but is no less valid regarding the seventeenth. "The Elizabethan world was populated, not only by tough seamen, hard-headed politicians, and serious theologians. It was a world of spirits, good and bad, fairies, demons, witches, ghosts, conjurors."[3] The world was one of magic and mankind needed to stay alert in order to overcome worldly temptation. "We cannot understand the motives and acts of our forefathers unless we take into consideration the mental condition engendered by the consciousness of this daily and hourly personal conflict with Satan."[4] Despite Roger Bacon's emerging scientific ideas, alchemy was classified by some thinkers as "*sept ars demonials*"[5] meaning that it was a process that needed Satan's help to succeed. James I seemed especially frightened of witchcraft resulting in the horrendous slaughter of women accused of sorcery.[6] Yet, this was the same time frame as the events that led to the creation of the Royal Society of London and the development of the scientific method. Cornelius Agrippa wrote about the distinction between witchcraft and magic. He explained that the word Magus, was a Persian word and that it referred to someone who studied divine processes. He quotes Plato claiming that the distinguished philosopher stated, "Magick is the art of worshipping God."[7] The world of spirit and the world of matter were yet to be separated.

I will begin by introducing a few of the participants in this part of Katherine Boyle's story who were not portrayed in the description of her work as a healer. It is a complex story and needs a great deal of social and political background that might seem surprising in a book about alchemy. Yet, I hope the reader will see the nexus between alchemy and apocalyptic thinking by the end of this chapter. First, I begin with Dorothy Moore, a dear friend of Katherine Boyle.

Dorothy Moore: Seventeenth Century Feminist

Dorothy Moore/Dury, née King lived circa 1612-1664. Her parents were Sir John King and Catherine Drury. Her father had been knighted in 1609 and granted land in County Roscommon and Sligo. In 1634, King became a Member of the Irish Parliament for Roscommon. Dorothy's family lived mostly in Dublin where it is believed she may have met her lifelong friend, Katherine Boyle Jones, Lady Ranelagh.[8]

Dorothy grew up in a home in which women's education was valued. She was extremely bright, reading Hebrew, Latin, French, and Greek. Her sisters, Margaret and Mary, also had a reputation for their intellectual prowess. This is in direct contrast to her friend, Katherine Boyle, whose father did not believe that women needed higher education.

Dorothy married Arthur Moore around 1620. It was not a pleasant marriage as Arthur was quite a drinker. He was the representative of County Armagh in the Irish Parliament at the same time as Sir John King, Dorothy's father, served in Parliament, the year 1634. Dorothy and Arthur had two sons, Charles and John. Arthur took part in the Protestant Reform movement which sent him to The Netherlands with Dorothy and the children. The boys were subsequently tutored by a well-known teacher in Utrecht, Voetius.[9]

By 1641, Dorothy was back in London and a widow. After the Civil Wars between Charles I and Parliament when England became a Commonwealth, Dorothy became quite politically active. One of her very close friends was Samuel Hartlib, a publisher, who worked at disseminating new ideas developing in Europe. He exchanged news, both religious and intellectual, with his correspondents on what was happening in England. This practice would eventually lay the foundation for the Royal Society's journal, *Philosophical Transactions*. Hartlib is an important associate for Dorothy in that alchemy was one of his significant interests. I will return to Hartlib in a moment. His story is fascinating, complex, and tangled in politics. Dorothy's surviving letters come mainly from the Hartlib collection.[10]

Eventually, Dorothy married her second husband, John Dury, a well-known Calvinist minister who was also quite active in politics. They had an interesting courtship in that they had been closely associated after her first husband's death, discussing and writing about various issues of concern. In due course, Dorothy realized that people were gossiping about the many hours she spent with John and pragmatically, decided that since she also needed a male colleague to more effectively disseminate her ideas in their patriarchal society, perhaps marriage was the best recourse.[11] Marriage was not an easy decision for Dorothy because she was concerned that by doing so she would be compromising her feminist beliefs in order to get her work accomplished. Lynette Hunter writes, "But working painstakingly through a theological discussion of election and 'calling' she logically deduces that marriage can be a fulfilment rather than an abdication."[12] Although Dorothy had previously taken the Moore surname, she was less willing to do so with John Dury. She signed her name as D.M. or D.M. Dury, as opposed to just Dorothy Dury. Dorothy's letters regarding her philosophical beliefs, that were written after her marriage in 1645, seem to have gone missing or were destroyed.

Dorothy wrote, *Of the Education of Girles*[13] sometime in the 1640s. The treatise may have begun in the form of a paper addressed to her dear friend, Katherine Boyle. Her husband, John Dury, wrote *The Reformed School* in 1650, in which he promoted women's education in science and languages as opposed to solely educating girls in needlecraft and dancing.[14] John and Dorothy were also deeply involved in the Millenarian movement that pertains to the alchemy-Apocalypse relationship that is the focus of this chapter.

Dorothy's circle of association included Princess Elizabeth of Bohemia as well as Anna Maria van Schurman (from Utrecht). Van Schurman (1607-1678) is described as a child prodigy who spoke 12 languages and excelled at philosophy, history, theology, geography, astronomy, painting, and poetry. Van Schurman, a formidable intellect, admired Dorothy and compared her to Lady Jane Grey, a very well educated woman who lived about 100 years prior. Van Schurman writes to Dorothy in a letter dated August 8, 1640:

> I am delighted to have heard about you and your reputation and thank heaven for knowing a lady like yourself, such a respected lady of your people, because owing to my ignorance, I thought that no clever woman had remained in England after the death of Jane Grey and Queen Elizabeth.[15]

Van Schurman's own writing tended to focus on women's equality, at least those from the upper class and their need to study science. Dorothy focused on theology and the need for educating girls from all social classes. From 1648-1650, Dorothy was ill from complications of childbirth. Her son, Charles, had died at a young age and the family experienced financial difficulty due to money owed them by the government. It is from these dire circumstances that we get a hint of Dorothy's alchemy. It is referenced in a letter written by William Hamilton, a friend, in which he writes, "'had thowght that Mistresse Durey . . . cowld farre lesse [could far less] have stooped . . . to have taken up a publick shop for selling of spirirts & oils.'"[16] Dorothy seemed to have ignored social mores that frowned upon her obtaining needed funds by selling chemicals and medication. After all, a mere vendor of medicines and ointments was considered lower class and not something in which a member of the upper class would participate. In another letter, John Dury wrote to Benjamin Worsley (member of the Hartlib circle) that he needed information on distillation requesting, "for 'vs, who are not acquainted with the practice.'"[17] Thus, it appears that Dorothy tried to supplement the family income by brewing medicines and healing tinctures, "women's alchemy."

Dorothy Moore gave birth to a daughter with John Dury in 1654, Dora Katharina. The child was sickly but managed to live. Due to the assassination of Charles I, John Dury, a Royalist supporter, had needed to move to Europe on a permanent basis. He remained living on the continent even after the Restoration of the monarchy in 1660 with Charles II being established on the throne.[18] Circa June 1664, Dorothy died at the age of about 52. Their dear friend, Henry Oldenburg, was given guardianship of Dora Katharina although John was still alive. Oldenburg's relationship with both Dorothy Moore, Katherine Jones, and their circle will be discussed more thoroughly in the following section. Dora Katharina was actually raised by a friend of Oldenburg's. Upon the death of

Oldenburg's first wife on August 13, 1668, he married Dora Katharina who was age 14. Years later, their children eventually became wards of Katherine Boyle and her brother Robert.

Katherine Boyle Jones, the Hartlib Circle and The Royal Society

The threads that bound the English upper classes were tangled and one finds relatives, spouses, and friends reappearing in various relationships when conducting historical research, somewhat like a convoluted plot in a mystery novel. Katherine Boyle Jones, Lady Ranelagh's circle of friends included the very same people who made up the Hartlib Circle, a group that has a fascinating history and an intriguing part to play in the alchemical story. Katherine was active in the Hartlib Circle and in fact, she and Dorothy Moore were frequent correspondents of Samuel Hartlib, the man I mentioned earlier, who corresponded with the European thinkers of the day. It was a rich intellectual community that included several women, increasingly even non-aristocratic women, up to the end of the 1650s. Many of the correspondents were also part of the Oxford circle that would eventually morph into the Royal Society. The participants were concerned with matters of natural philosophy and included Robert Boyle, Thomas Willis and John Wilkins.

Katherine was involved with Samuel Hartlib in the context of having her residential address on Queen Street serve as the clearinghouse for the correspondence that would eventually lead to the creation of the *Philosophical Transactions of the Royal Society*. A brief discussion of the Royal Society will be useful because it was such a dynamic element of the era about which I am writing and still continues to play an important role in scientific advancement. The history of the creation of the Society may seem overly encyclopedic in light of the focus of this chapter. However, the participants in this story are completely intertwined with the Boyles. This brief history also gives some sense of the place that alchemy still had in scientific thinking.

The Royal Society's beginnings are a bit murky but it seems to have been an evolution of two study groups that met prior to 1662, in Oxford and at Gresham College in London.[19] Followers of Francis Bacon's theories on scientific experimentation met weekly to discuss his methods. Participants were not limited to what we might think of as traditional scientists, but included physicians, architects, poets, members of the clergy, lawyers, and government officers. For example, one of the future members and secretary of the Royal Society was Heinrich (Henry) Oldenburg, a German scholar who resided in London. Oldenburg became the Secretary of the Royal Society of London in 1665, as well as editor of the *Philosophical Transactions*. He was born in Bremen circa 1617-1620 and became well versed in the sciences having a father who was a physician.[20] Oldenburg completed a Master of Theology in 1639 and then seems to have worked as a tutor while learning English, French, and Italian, skills that would serve him well. In 1653 Oldenburg was appointed an envoy to England. He decided to remain there and met Katherine Boyle Jones who was now Lady Ranelagh. He tutored her son, Richard Jones. Oldenburg also met Samuel Hartlib of the alchemically minded Hartlib Circle. Thus, the reader can see how interconnected these groups were and how difficult it is to separate their activities and beliefs into distinct categories.

Oldenburg studied at Oxford at the same time as John Wilkins, John Wallis, and Jonathon Goddard whose meetings were known as the "Oxford scientific 'company' or 'club.'" [21] The Oxford group was also known as the "Invisible College."[22] This group became associated with another study group at Gresham College in London which eventually led to the founding of the Royal Society. The Oxford group was more focused on chemistry and medicine whereas the Gresham scholars emphasized math and astronomy. By 1660, the Gresham meetings were held regularly and one could subscribe to the group for one schilling. The members resolved to form a scientific society in 1660 but this was not accomplished formally until King Charles II granted them a charter in July 1662. It might interest the reader to know that Charles II was also interested in alchemy and kept a laboratory at his palace in Whitehall.[23] Thompson quotes Samuel Pepys' diary entry from January 15, 1668 in which he wrote that he had seen "'the King's Elaboratory underneath his closet. A pretty place, and there saw a great many chemical glasses and things, but understood none of them.'"[24] The same Pepys who wrote about Margaret Cavendish was a British government official who kept an amazingly detailed diary for nine years, 1660-1669, that is a significant window into that period of time. "King Charles' Drops"[25] became a well-known medicine created by the monarch.

Original members of the Royal Society included Robert Boyle, Robert Moray, Christopher Wren, and John Wallis. Subsequently, Isaac Newton, Christian Huygens, Robert Hooke, and Henry Oldenburg became members. In 1663 the Royal Society was granted arms and their Latin motto read "And do not ask, by chance, what leader I follow or what godhead guards me. I am not found to revere the word of any particular master."[26] The Royal Society's journal, *Philosophical Transactions* began publication in 1665. It remains a scientific institution to this day.

The first volume of the *Transactions*, covering the years 1665 and 1666, is titled, *Philosophical Transactions: Giving Some Accompt of the Present Undertakings, Studies, and Labours of the Ingenious In Many Considerable Parts of the World*. Henry Oldenburg, the editor, wrote a wonderful introductory remark that ends with his offering the work to the reader declaring, "The Great God prosper You in the Noble Engagement of Dispersing the true Lustre of his Glorious Works, and the Happy Inventions of obliging Men all over the World, to the General Benefit of Mankind: So wishes with real Affections."[27] The *Transactions* reported what was being discovered in the areas of early chemistry, optics, physics, mathematics, agriculture, mining, physiology, and astronomy. The contributors described phenomena that the modern reader might call fantastic but the author did so with scientific sincerity. One example is a report titled, "Some anatomical observations of milk found in veins, instead of blood."[28] Another report was, "Of a place in England, where, without petrifying water, wood is turned into stone."[29]

Oldenburg was not a practicing scientist but a student of the sciences making him an excellent Secretary for the Society and editor of the *Transactions*. He also became Robert Boyle's literary agent and seems to have had some experience dabbling in alchemy. He relates in a letter that, "I took one grain of it [vitriol] and projected it upon an ounce and a quarter of mercury, which I changed into the purest silver."[30] In a letter to Hartlib dated June 25, 1659, Oldenburg again writes concerning alchemical work.

I send you here enclosed a Chymicall process of vitrioll (in acknowledgment of ye [the] secret you sent me, wch [which] shall not loose the name of a secret for me) and intreat you, to communicate it to none, but noble Mr. Boyle, who, I am sure, upon my desire will impart it to none but My Lady Ranalaugh, [Katherine Boyle Jones] wch is a person, yt can keep a secret as well, and any I know.[31]

Robert Boyle's work at Katherine's home in Pall Mall and his association with the Hartlib Circle were closely connected to the changes taking place in the new science. A disagreement had emerged with Descartes who supported a deductive methodology as opposed to the inductive methodology of the new scientific theory set forth by Sir Francis Bacon. Although Boyle studied alchemy extensively, he wrote in *The Skeptical Chemist*, that alchemy had become old school and the new chemistry was the way to proceed.[32] However, that may reflect the variations in the way historians have interpreted *The Skeptical Chemist*. Lawrence Principe argues that Boyle did not negate alchemy and that those reports have been exaggerated by historians. Boyle maintained a belief in the Philosopher's Stone but not in a Paracelsian theoretical framework.[33] The way to produce the Stone was more the issue, not its existence.

Theology was never far away from the new scientific discoveries. The juxtaposition of science and theology is evident in Katherine's thinking about the meaning of the Plague as well as the Great Fire of London. Katherine Boyle wrote to her brother, Robert, on September 9, 1665, during the height of the Plague, a time that seemed to be a sign of the Apocalypse to Millenarians. She referred to a publication titled *Bills of Mortality* that listed the names of those who had died and wondered about the connection to the End Times. Writing again, a year later, on September 12, 1666, Katherine discussed the ruin of London after the Great Fire (September 2, 1666), another possible harbinger of the End Times, especially having taken place in the year 1666, one of the predicted years of the Apocalypse. In another letter dated, November 13, 1666, Katherine related a story that was being told about how after the Fire, the corpse of Thomas, Bishop of London (1402) had been discovered buried and in good shape except for being dried out. Katherine suggested that since the ground in which he was found was chalky and probably limestone, that was most likely what had kept him so well preserved. Her scientific thinking was a natural part of her observations and not separated from her theology or daily life. On September 11, 1677, Katherine is expressing her sadness to Robert over the deaths of Henry Oldenburg and Dr. Worsley that came close to each other. She articulated how better it was to grieve openly than bottle things up:

> My experience . . . has taught me that its safer to have those uneasy things to us soe farr touched upon as to beget some vent for such sorows rather than by smothering them within our selves to continue to us a longer exercise under them.[34]

Katherine also employs words of comfort that allude to her friends being in Heaven, a better place, and where she and Robert as well as other pious folks hope to go.

The women practicing alchemy were not so quick to totally dissolve their association with the old ways. They continued to brew their medicines, experiment with chemistry, and write about philosophy as we have seen with Anne Conway, Alathea Talbot, and Margaret Cavendish. Some of the debate is as much about the history of women as it is the history of science. It is about the continuing dance between the feminine and the masculine. It requires a different understanding about how their work is to be defined and redefined—and in the same way as alchemical matter, the operations must be revisited and refined over and over again.

The Millenarian Movement

In the alchemical operation of *separatio*, the matter being worked upon must be separated into its basic elements and purified before they can be reunited in perfection. I have found that as I researched the women alchemists, it was easy to become confused by the way their politics, science, and theology were interconnected. What was the significant element of their story and did any of it explain their relationship with alchemy? Therefore, I have attempted to look at the areas of theology, science and politics individually in an attempt to reunite them into a coherent picture of the women alchemists' work. In this section I will look more closely at their theology.

Katherine Boyle Jones was both pious as well as tolerant regarding religious issues. Like her best friends, Dorothy Moore and John Dury, her inclination was toward millenarianism, a belief regarding Christ's return to earth according to the prophecy described in Revelation. Many Christians believed that just as God created the earth and was incarnated as Jesus on earth, that the last days of judgment would also take place, not in heaven but on earth.[35] Defining the term Millennialists or millenarian requires one to take into account two main points of view regarding the Apocalypse. Pre-Millennialists believed that Christ would make a second appearance *preceding* the end of world to be followed by 1000 years, a millennium, of peace and harmony. Post-Millennialists believed the world would consist of 1000 years of war and desolation, followed by Christ's second coming. There are also sub-groups who argued in favor of further explanations of when Christ would make an appearance that get more and more convoluted as one attempts to follow their reasoning.

The common thread held by the different Christian points of view were that one could look at world events and determine how they were leading up to the Second Coming of Christ. The books of Daniel, Luke, and Revelation were the main sources of information concerning the events that would precede the Apocalypse or End Times. The dates of the Second Coming have typically coincided with times in history when life was difficult for the poor or dispossessed; the promise that better times were coming was both hoped for and expected. However, around the fifth century CE, the Catholic church began teaching that Revelation and Daniel were about *spiritual* enlightenment as opposed to a concrete, worldly happening that could possibly destabilize its power.[36] The church could not conceive of being overthrown by some unknown power, nor would it succumb to its annihilation easily. This is ironic in that one believed that what happens must be God's will. The church was not going to accept that God's will ran counter to theirs.

The Kabbalistic Connection with Christian Theology

The millenarianism practiced by Dorothy Moore and Katherine Boyle Jones included their study of the Kabbalah as well as other treatises of alchemical and/or occult practice. James Jacob argues that there was a clear association between millenarians and devotees of the alchemist/medical doctor, Paracelsus. He further contends that part of the context of the uprising against Charles I was that the monarchy monopolized industry, the law, and medicine. Upon abolishing the monarchy, more common medical practitioners who followed Paracelsus such as surgeons and apothecaries, who were also unable to be accepted into the Royal College of Physicians, could emerge in a stronger position.[37] For example, Samuel Hartlib, a millenarian, was also involved with translating Paracelsus' work into English. How the two are linked is unclear; however, alchemists commonly studied each of these fields.

In another letter written by Anna Maria van Schurman to Dorothy Moore dated April 1, 1641, we see the exchange of ideas concerning Pico della Mirandola (1463-1494). Pico was a Renaissance philosopher and Christian Kabbalist associated with the Medici's in Florence.[38] Pico wrote *Opera omnia* containing *72 Kabbalist Conclusions*. The original *Kabbalistic Conclusions* first appeared in Rome in 1486 but that edition only contains 49 of the conclusions leading Arthur Waite to note that the rest are attributed to Pico.[39] The *Conclusions* address Creation, Adam and other brief but important statements regarding the universe and our place in it. According to the 13th conclusion, "He who shall know the Mystery of the Gates of Understanding in the Kabbalah shall know also the mystery of the Great Jubilee."[40] Pico grounded his thinking in antiquity, focusing on treatises such as the *Corpus Hermeticum* attributed to the great alchemist, Hermes Trismegistus.

We see in the thinking of Pico (as well as Isaac Newton), and women such as van Schurman and Dorothy Moore, a belief in *prisca sapientiae*, or the wisdom of the ancients. These old alchemical texts were thought to have been written as early as the time of Moses and were considered sacred. The study of Kabbalah by the alchemists clearly connects their practice to the more spiritual aspects of the work.

> The Hebrew Cabalists believed that their teachings went right back to Moses through a secret doctrine which had been handed down through initiates. And since for Pico, Cabala confirmed the truth of Christianity, he believed it to be a Hebrew-Christian source of ancient wisdom which corroborated not only Christianity, but the Gentile ancient wisdoms which he admired, particularly the writings of 'Hermes Trismegistus.' Thus, Christian Cabala is really a keystone in the edifice of Renaissance thought on its 'occult' side through which it has most important connections with the history of religion in the period.[41]

For example, Johannes Reuchlin published the first book on Kabbalah by a Christian scholar and Gentile in 1494.[42] The reader will remember from chapter two that Kabbalah refers to the framework for Jewish mysticism in which the world is made manifest from God, who is eternal and without limits.[43] The alchemical parallel is the *prima materia* discussed earlier and the way the world is manifested out of that first substance. The *prima materia* of Kabbalah contains all that there is in the universe which is of God and is manifested as our world. Thus, the motif of the world being born out of chaos is seen in both Kabbalah and in alchemy.

The interconnectedness of matter is also described and is illustrated by the Great Tree of Azilut that is a concretization of Kabbalistic thinking.[44] The nomenclature might vary throughout time but the functions of each Sephirot did not. Moreover, the Great Tree bespeaks balance in that the sides balance each other. There is a passive pillar balanced by the active pillar. The feminine aspect is found on the passive side whereas the active pillar contains the masculine aspect, much like the ideas of yin and yang. (See Figure 7.1, next page)

Halevi postulates that one's temperament guided one's approach to studying Kabbalah. It could be taken literally, read as an allegory, thought of as a metaphysical framework, or experienced in a mystical way. He cautions that one does not approach the study of Kabbalah lightly, warning: "To play with the Truth is not encouraged, because the first stages of initiation are concerned with the image one has of oneself, and few can withstand the exposure of their illusion."[45]

Passive Pillar   Keter   *Active Pillar*

Binah   Hochma

Daat   Keter divides into
Binah and Hok-
mah—Eternal
Gevura   Chesed   Divine/ Holy of
Holies.

Tipheret

The Shekinah
or Divine Pres-   Hod   Netzah
ence dwells in
Malkhut.   Yesod

Malkhut

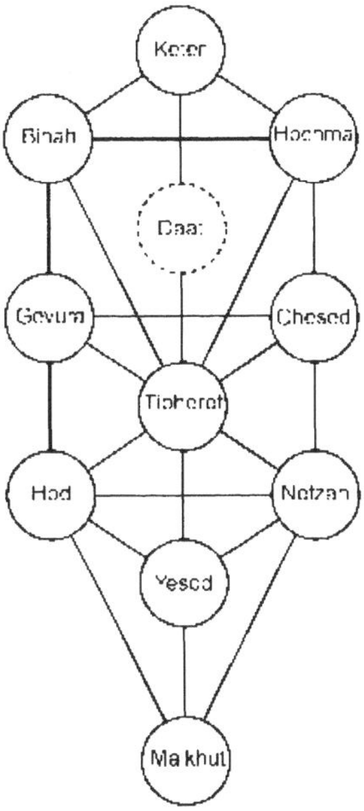

Figure 7.1: The Great Tree of Azilut with Halevi's Exposition

It should be no surprise that the alchemists studied this Jewish mysticism; it paralleled their work so well. The world becomes manifest from God and if one can understand that framework, one can begin to comprehend the nature of God. Millenarians also believed that Jewish mysticism held truths that were vital to their beliefs. For example, many Christian Kabbalists viewed the Kabbalah as a support for their Christian doctrines and were convinced that it provided the link between Jewish and Christian thinking.[46] The name of God serves as another helpful example of the nexus of Christian-Kabbalistic thinking.

One of the basic Christian-Kabbalistic proofs of Christ's divinity was found in the example of the name of Jesus and the Tetragrammaton. The Tetragrammaton is the four-letter sequence from the Hebrew alphabet Yod - Heh - Vav – Heh, which we pronounce YAHVEH or Yahweh.[47] God revealed the name to Moses; however, since the Tetragrammaton was believed to be unpronounceable, one articulated out loud the word *Adonai* (the Lord), or *ha Shem* (the Word).[48] In Hebrew the name looks like: יהוה where י = yohd; ה = heh; ה = vahv; and finally, ה = heh are read from right to left. "Yod" is father, "He" translates to mean son, and "Vau" is the Holy Spirit. Godwin shows a diagram in which the Tetragrammaton is also written vertically with Yod on top, then descending from He to Vau to He. Godwin also illustrates the link between

Kabbalah and Christian esoteric beliefs when she quotes from the *Fama* (a Rosi-crucian treatise) " 'sub umbra alarum tuarum JEHOVA' meaning: "above and beyond all is Yod, the letter from which all proceeds and which conceals in it-self the whole Name."[49] Halevi notes that this intriguing picture of the Hebrew letters represents a picture of a man. That man is Adam or, first man.

Figure 7.2: The Tetragrammaton as Adam Kadmon

In Hebrew, the name Jesus is written in the same way as the name of God (Tetragrammaton), with a letter inserted in the middle, *sin*. Thus, we have the Tetragrammaton, that is the name of God that cannot be spoken. Yet, by adding this middle letter, the name Jesus, the name can now be spoken aloud and thus, Jesus coming into the world allows us to hear the name of God.

Figure 7.3: The Pronounceable Name of God

Yates writes, "the S in the Name of Jesus makes audible the ineffable Name (composed only of vowel sounds) and signifies the Incarnation, the Word made flesh or made audible."[50]

Believers such as Dorothy Moore, Katherine Boyle, and their circle hoped that their study of the Kabbalah would prove to many Jews that Jesus was truly the Son of God, supporting their Christian position. Matthew and Harrison write in the *Oxford Dictionary of National Biography* regarding their interest in Juda-ism:

> was also connected to the same millenarian and tolerationist interest in Judaism that led her [Katherine] and Robert Boyle to invite the Dutch Jewish leader, Manasseh ben Israel, to meet them during his visit to London in 1656, on his mission to persuade Oliver Cromwell to permit Jews to settle in England.[51]

Millenarians such as Dorothy Moore and Katherine Boyle, were fully aware of the part of the prophecy that stated that the Second Coming would be preced-ed by the union of Christians and Jews. They understood the Bible to mean that Jews would need to be converted to believing that Christ was the only Son of

God for the 1000 year cycle of the Second Coming to begin. This necessary union created the impetus for allowing expelled Jews back into England.

John Dury was interested in rabbinical teaching and knew a number of rabbis on the continent which was how he became acquainted with Mannaseh ben Israel. John Dury believed (probably unbeknownst to the rabbi) that ben Israel would be able to help convert Jews to Christianity as a prequel to the beginning of the millennium.[52] Dury and Samuel Hartlib also hoped to build a university for Jewish studies as yet another means of preparing for the events in Revelation.[53] Manasseh ben Israel believed in the coming millennium; however, he was expecting the original or First Coming of the Messiah. In a letter written to Manasseh ben Israel by Henry Oldenburg on July 25, 1657 he states, "Relating to the coming of the Messiah, equally desired by you and by us. I say the coming, meaning that which you take to be the first and we are persuaded will be the second.[54]

Yet, as I mentioned, anticipation of the Apocalypse became a contradiction to the teachings of the Catholic church which argued that the ideas in Revelation and Daniel were metaphors for the life of the church and that the 1000 year period would begin from the time of "change from sin to purity."[55] The timing of that change is unclear.

Fifth Monarchy Men

The millenarian view was shared by not only the Hartlib Circle, but also other groups such as the Fifth Monarchy Men. The Fifth Monarchy Men's story further illustrates the context of the seventeenth century with its intertwining of science, politics, and theology. I hope the reader can see how making sense of the study of alchemy is only truly understood by talking about the times in which the practitioners lived and pursued their goals. The Hartlib Circle was not part of the Fifth Monarchy movement; however, they did hold similar aims. Henry Archer was a fellow who preached about the coming of the fifth monarchy or Millennium.[56] The first four monarchies were described in Revelation as Babylon/Assyria, Mede/Persia, Greece, and the current Roman Empire. The four monarchies having passed would be followed by the fifth, which would begin with the reign of Christ. They believed that Christ would return to earth, set up his kingdom, depart again to heaven for 1000 years, and finally, come back to earth to carry out the Day of Judgment. Thus, in this framework, there was a kind of third coming where the sinners were to be punished.

Calculating the date of the Second Coming entailed some interesting reasoning. Archer published his conclusion in a book, *The Personall Reign of Christ on Earth*.[57] If the reader has not read Revelation, I advise you to do so as it is both interesting and a useful foundation for understanding Archer's ideas. One of the apocalyptic beasts in Revelation is described as having 10 horns which are said to represent the 10 kingdoms that had emerged as a result of the decline of the western Roman Empire. The creature also sported a small horn which signified the papacy to many Protestant interpreters. In the prophecy, the little horn was to be destroyed after "a time, and times, and half a time" (Rev. 12:14). Revelation 10:2 refers to a city that would be held captive for 42 months along with two witnesses who would have the power to prophesy for 1260 days. The 42 months is equivalent to 1260 days if one multiplies 42 (months) and 30 (days). However, a Biblical day was thought to actually be equal to one year; thus, the countdown to the second coming would be 1260 *years* from another calculated beginning point. Archer agreed on 400-406 CE as the time when the 10 kingdoms emerged from the Roman Empire and the Papacy began its rule. He argued that the Bishop of Rome or the Pope, began usurping powers that Archer believed were not legitimate. For some reason, Archer added 406 years to the 1260 and concluded the Second Coming would occur in 1666 which was very significant as 666 is the number of the beast. "Here is the wisdom. Let him that hath understanding count the number of the beast: for it is the number of a man; and his number is Six hundred threescore and six" (Revelation 13:18, KJV). Archer's arguments sound fantastic to the modern ear but his calculations were taken seriously by millenarians. Recall that the occurrence of the Great Fire of London in the same year, 1666, gave many such as Katherine Boyle, pause to consider if the End Times were near.

Archer employed a second method of calculation using the year when the Jews would be converted to Christianity as the initial countdown point. He cited Daniel 12:2 which declares, "And many of them that sleep in the dust of the earth shall awake, some to everlasting life, and some to shame *and* everlasting contempt"(KJV). In other words, the Jews were perceived to be asleep to the Truth and would either awaken to the Truth or be doomed for eternity. The year 360 or 366 was calculated as the time when the Roman Emperor, Julian the Apostate, instituted pagan practices such as labeled as "Heathenisme" meaning they were associated with the heathens. When Archer added 1290 to that date, he concluded that either 1650 or 1656 were the dates when the Jews would be converted to Christianity. Conversion did not appear to be a joyful event for the Jews as Archer claimed they would suffer for 45 years. Christ would finally appear, followed by 1000 years of his kingdom. As a consequence Archer concluded, "'it is likely that Christ's coming from Heaven, and raysing the dead, and beginning his kingdome, and the thousand years, will bee about the yeare of our Lord 1700, for it is to be about fortie-five yeares after 1650 or 1656.'"[58] There is a convoluted logic to these confusing calculations but let the reader be aware, although this thinking sounds irrational, even Newton had his own calculation for the Apocalypse which was to occur around 2060.

Regardless of the dates proposed, Phillip Rogers describes how the Fifth Monarchy Men focused their energy on making preparations for the coming of Christ. They believed that the rule of Charles I was preventing the hoped for ending of the fourth monarchy and the eagerly anticipated beginning of the fifth. One of their members, Colonel Thomas Harrison, emerged as both an enemy of Charles I during the Civil Wars and a millenarian. Many officers in the British army, along with Harrison, succumbed to these beliefs and therefore, saw themselves as not just solely an army for England but as the literal army that would destroy the antichrist.

Although Oliver Cromwell was the leader of the coup against Charles I, unlike the Fifth Monarchy Men, he did not support dealing with Charles as a criminal. However, it is believed that Harrison eventually prevailed and Charles was tried and executed on January 30, 1649, paving the way for the return of Christ. The Fifth Monarchy Men continued to manipulate Cromwell and the government to assure that Christ, or King Jesus, would be set to rule.[59]

John Rogers, a significant member of the Fifth Monarchy Men, emerged as an ally for Cromwell. His childhood experiences led him to be terrified by Puritan extremism, resulting in his developing a complex set of compulsive rituals that included memorizing and reciting sermons, sleeping with his hands in a prayerful position, and constantly being on watch for any behavior that would result in him being damned to hell.[60] He was a perfect candidate for the Fifth Monarchy Men.

As Cromwell dissolved the Rump Parliament in 1653 and created the new Commonwealth, the Fifth Monarchy Men were there, giving advice and admonishing him to be sure that the way stayed clear for Christ and his saints whose identities were still unknown. They planned to create an assembly of moral and righteous men who would be elected from their churches. The newly established assembly met on July 4, 1653 and began by instituting many of the Fifth Monarchy Men's values regarding the courts, rules for tithing, and how to care for the poor. However, moderates were concerned with this political-theological government and Sir Charles Wolseley met with Cromwell to sort it out. The moderates decided to place all the power of the assembly under Cromwell in order to dilute the Fifth Monarchy Men's power.

The new Parliament was thus instituted and the Fifth Monarchy Men were not pleased. Two more Fifth Monarchy Men emerged as fanatic radicals, Christopher Feake and Vavasor Powell. They implied in their harangues that rather than Charles I having been the Little Horn on the Beast of the Apocalypse, it was actually Cromwell. They even turned against the army in light of its allegiance to Cromwell. The government gave the Fifth Monarchy Men many warnings and made arrests resulting in Harrison losing his Army commission. Although the leaders of the Fifth Monarchy Men were being arrested, they preached sedition, fanatically convinced that they were doing the work of the Lord. Insurrection was planned but not carried out. Subsequently, Cromwell died and his son, Richard, succeeded him. After continued problems and worries of anarchy, Parliament entreated Charles II to come home and to restore the monarchy. He was proclaimed king on May 8, 1660, a terrible shock for the Fifth Monarchy Men.

Harrison was arrested, tried, and executed on October 13, 1660 as the instigator of regicide of Charles I. To the very end, Harrison behaved as a martyr for the cause. John Carew, a fellow member of the Fifth Monarchy Men, also suffered the same fate. Their martyrdom set the stage for the fanatical Fifth Monarchy Men leader, Thomas Venner, to lead a last ditch uprising. There was fighting over a period of a few days but most of the fifty Fifth Monarchy Men were either wounded or killed. However, they managed to kill a lot of the "trainband"[61] or civilian militia. Venner was eventually executed for treason January 1666.

Consequently, we can see how life for these early thinkers who are the subject of this book, was not black and white. One might be a theologian but this could not be divorced from politics or science if one wanted to be taken seriously. Even George Fox, the well-known leader of the Quakers illustrates the intertwining of politics and theology.[62] In a letter he wrote in 1661, George Fox described a previous conversation he had with Jesuits where he had questioned the purity of the church. He asked them, "whether ye Church of Rome was not degenerated from ye Church in ye primitive times from ye spiritt & power & practice yt they was in ye Apostles time."[63] The Jesuits did not give him a satisfactory answer and he accused them of thinking that they were as moral as the Apostles.

Fox moreover, did not appreciate (along with numerous other reform movements) the Catholic practice of praying to images and putting people to death for their religious beliefs, although George Fox may have needed to be reminded of Thomas Cromwell's earlier use of the rack on Catholics during the time of Henry VIII. Fox talked about how the Quakers had argued with other sects such as the Baptists, Lutherans, Calvinists, Arrians, and Fifth Monarchy Men. The tone is fierce as can be heard as one reads a small excerpt. Notice the fight paragraph where Fox refers to the Fifth Monarchy Men members as "beasts and whores."

> And ye 5th Monarchy men I was moved to give foorth a paper to ym[them]: whoe lookt for Christ personall comeinge in :66: & some of ym did prepare ymselves when it thundered & rained & thought Christ was comeinge to sett uppe his kingedome & then they thought they was to kill ye whore without ym but I tolde ym ye whore was alife in ym & was not burnt with Gods fire: & Judged in ym: with ye same power & spiritt ye Apostles was in.
>
> And they lookt for Christs comeinge to sett uppe his kingedome & there lookeinge was like unto ye pharisees loe heere loe there: but Christ was come & had sett uppe his kingedome above 1600 yeers since: accordinge to Nebuchadnesars dreame & Daniel prophesy: & hee had dasht to pieces ye 4 monarchys & ye great image with itts heade of golde & sylver breast: & belly of brasse: & Iron leggs: & feet part Iron & part clay.
>
> And therfore all yee 5th Monarchy men yt bee fighters with carnall weapons: yee are none of Christs servants but ye beasts & whores. And Christ saith all power in heaven & earth is given to mee soe then his kingedome was sett uppe & hee reignes: & wee see Jesus reigne sayd ye Apostle: & hee shall reigne till all thinges bee put under hie feete [his feet] though all thinges is not yett putt under his feete nor subdued.[64]

The Quakers were not immune to criticism and may have not been entirely truthful as evidenced by some opinion. Lord Conway, husband of Anne Conway wrote to his brother on July 5, 1659, accusing Sir Harry Vane, Humphrey Salway, and the Quakers of hypocrisy writing that their "design is only to turn out the landlords"[65] and appropriating the land is implied.

Clavis Apocalyptica

The circle in which Dorothy Moore and Katherine Boyle Jones studied, did not, as I have mentioned, subscribe to the radicalism of the Fifth Monarchy Men. However, they did analyze another apocalyptic book titled the *Clavis Apocalyptica*. John Evelyn, one of the Circle, wrote in his diary on April 26, 1689 about a conversation that he had with a Bishop. They believed that the third trumpet signaling the Apocalypse had sounded and that the third vial had been poured out. He references Joseph Mede, the author of the *Clavis*, stating that they agree with his interpretations regarding the book of Revelation.[66]

The *Clavis,* or key, was written by Joseph Mede who was born in 1586 in Essex. He graduated from Cambridge and became well-known for his careful scholarship. Mede also subscribed to the belief that the resurrection of the saints and martyrs would occur as an event preceding the advent of the millennium.[67] The translator of the *Clavis*, Richard More, related that after reading the *Clavis*, he was so taken by it that he decided to translate it from the Latin.[68] More included references to Biblical verses in the margins of the translation for the ease of the reader. At the time that the book was in press, More related a story where a gentleman, Master Haydock was supposed to have contacted him and shown him letters that he claimed to have exchanged with the original author, Joseph Mede. It seems that Haydock wanted More to print the book in a way that both he and Mede thought best, implying that they were well aware of and approved, the upcoming translation prior to its publication. Mede and Haydock hoped that the *Clavis* would be published in the form of seven separate, sealed parchments that would be opened one after another as is described in Revelation. The *Clavis* contains Mede's comments and interpretations of Revelation as well as his own calculation of the second coming.

The photo of the wished for layout shows a large scroll in the form of seven sections, each attached to a leather strap that when closed, gives the book the look of having seven straps encircling the scroll. Each leather strap fastens shut with the idea that the reader would open each section one at a time, in order, in essence, to reenact the prophecy of opening the *Seven Seals of Revelation*. The scroll is being handed to mankind by the *Hand of God*. Interestingly, the illustration shows Book Four opened for viewing with a fellow hung by his right foot upside down, exactly as ones sees in the Hanged Man card in Tarot. There is also a man pictured who is hanging on a cross upside down, possibly referring to Saint Peter. The mixing of the occult with theology is well demonstrated.

Another diary entry included by John Evelyn shows that Katherine and Robert were still active in the Millenarian movement as late as 1690, just a few years prior to their death. On June 18, 1690 John writes about visiting Robert and Katherine with the Bishop of St. Asaph to speak with them about the state of the world. Conditions in France seem to portend that the fall of the King of France would be followed by the return of the Jews to England (a necessary precursor to the Second Coming). He goes on to declare:

> The Kingdom of the Antichrist would not be utterly destroy'd, till 30 years, when Christ should begin the Millennium, not as personally and visibly reigning on earth, but that the true religion and universal peace should obtain thro' all the world. . . . Mr. Brightman, Mr. Mede, and other interpreters of these events fail'd, by mistaking and reckoning the yeare as the Latines and others did, to consist of the present calculation, so many days to the yeare, whereas the Apocalypse reckons after the Persian account, as Daniel did, whose visions St. John all along explains as meaning the only Christian Church.[69]

To my modern mind, it would seem that the coming and going of years in which the Second Coming was predicted would be rather disheartening. Perhaps it was not so for the Millenarians who seemed to hold to their faith and conclude that the calculations were from human error, not from faulty theology.

Placing the Millenarian Movement in a Depth Psychological Perspective

I include this background information regarding Dorothy Moore's and Katherine Boyle's philosophical and theological studies as it begins to illuminate the part that alchemy may have played in their lives, as well as for many of the other *sorores mystica* examined in this book. Hannah discusses Jung's work on the *Mysterium Coniunctionis*, noting that Jung "emphasized that how far the alchemist succeeded in his endeavors is really much less important than the fact that he was gripped by the numinous archetype behind his effort, so that he went on trying without interruption throughout his whole life."[70] I stated in the Introduction that I was not solely interested in the names of women who practiced alchemy but as well, wished to understand their lives and why they did that work. Theology seems to be a critical paradigm for many of the women alchemists I have researched thus far. They focused on the world of the numinous and in their day, that would have taken the path of theological thought.

This is consistent with their alchemical brothers who were very focused on God's concrete role in the alchemical operations and in their belief that they needed to be worthy to partake in the work. God was ruler of the universe and thus logically, all nature including alchemy specifically, reflected him in some way. Carrying out the alchemical operations was a sacred act in a way that was very different from later chemistry. One approached alchemy with piety and respect for all that it symbolized. God was never far away from the work which makes it easy to see how studying theology and alchemy could be reconciled.

What did alchemy reveal about God? It would seem that God worked in the natural world with great complexity but the ultimate goal was perfection of matter as well as spirit. If Christ was the embodiment of God in matter, then alchemy was a window into the Trinity. Just as God the Father, Son, and Holy Spirit were considered to be One or perfection, so must the flawed human strive for perfection in him or herself. That perfection was found in the operation of *coniunctio*, the merging operation that produced the Philosophical Child, the Philosopher's Stone, the Christ image. Thus, alchemy became both a practical and spiritual practice.

Earlier, I referred to the dearth of written evidence concerning women's alchemy and would like to suggest that their theological activities are just that. The quote from the van Schurman letter (April 1, 1641) referred to previously, suggests that some women of the seventeenth century intuitively understood how illuminating it could be to recognize contrasting modes of thinking in *relationship* to one another, as opposed to a black and white rejection of one position over another. In their theology, it was not only revealed in the Bible but made sense that the Christians and Jews must come together for God's plan to proceed. In a sense, this concretized the operation of *coniunctio*. These women continued to struggle with the idea that spirit and body were interconnected. It then seems logical that a theology that dealt with matters of the spirit would be intertwined with their study of the natural world.

Notes

1. Joseph Mede, *Clavis Apocalyptica (English) or The key of the Revelation, searched and demonstrated out of the naturall and proper charecters of the visions. With a coment thereupon, according to the rule of the same key, published in Latine by the profoundly learned Master Joseph Mede B.D. late fellow of Christs College in Cambridge, for their use to whom God hath given a love and desire of knowing and searching into that admirable prophecie*. Translated into English by Richard More of Linley in the Countie of Salop. Esquire, one of the Burgesses in this present convention of Parliament. With a praeface written by Dr Twisse now prolocutor in the present Assembly of Divines (Imprint Printed at London : by R.B. for Phil. Stephens, at his shop in Pauls Church-yard at the signe of the gilded Lion, 1643) 23-27.

2. Theodore.Kirkingius, *Basil Valentine: His Triumphant Chariot of Antimony, with Annotations of Theodore Kirkingius, M.D.* (London: Dorman Newman at the Kings Arms in the Poultry, 1678), 79.

3. Frances Yates, *The Occult Philosophy in the Elizabethan Age* (London: Routledge,1979), 87.

4. Henry C. Lea, *A History of the Inquisition of the Middle Ages, 1-3* (New York: Harper and Brothers, 1888), 382.

5. Lea, 436.

6. Mitchell Salem Fisher, *Robert Boyle: Devout Naturalist, A Study in Science and Religion in the Seventeenth Century* (Philadelphia: Oshiver Studio Press, 1945).

7. Henry Cornelius Agrippa, *Fourth Book of Occult Philosophy*, Robert Turner, trans. (London: Askin Publishers, 1978/1655), A3.

8. Hunter (2004).

9. Voetius was well known in the Netherlands. He taught at the University of Utrecht, disagreed vehemently with Rene Descartes, and worked tirelessly to uphold the Reformist Movement.

10. Lynette Hunter, *The Letters of Dorothy Moore, 1612-64: the Friendships, Marriage and Intellectual Life of a Seventeenth-century Woman* (Aldershot, Hants: Ashgate, 2004); James R. Jacob, *The Scientific Revolution* (New York: Humanity, 1998).

11. Hunter, 2004

12. Hunter, xvi.

13. Dorothy Moore, *Of the Education of Girles* (c. 1650).

14. John Dury, wrote, *The Reformed School*, in 1650.

15. Hunter, 1.

16. Hunter, xxii.

17. Hunter, xxii.

18. Jacob (1998).

19. Edward N. da Costa Andrade, *A Brief History of the Royal Society* (London: The Royal Society, 1960); Rupert A.., Hall and Marie B. Hall, *The History of the Royal Society of London*, Vol. 1-4 (Hall and Hall, 1965); John Wallis, *A Defense of the Royal Society, and the Philosophical Transactions, particulary Those of July 1670* (London: for Thomas Moore, at the Maidenhead, 1678).

20. Hall and Hall, (1965, volume 1).

21. Hall and Hall, xxxvi.

22. Manly P. Hall, *Orders of Universal Reformation: Utopia* (Los Angeles: The Philosophical Society, 1949).

23. C. J. S. Thompson, *Alchemy and Alchemists* (New York: Dover, 1932/2002).

24. Thompson, 146.

25. Thompson, 146.

26. Andrade, 4.

27. *Philosophical Transactions: Giving Some Accompt of the Present Undertakings, Studies, and Labours of the Ingenious In Many Considerable Parts of the World* (London: Royal Society of London),1.

28. *Philosophical Transactions* (1665-1666), 100.

29. *Philosophical Transactions* (1665), 101.

30. Hall and Hall, 276.

31. Hall and Hall, 270.

32. R. H. Fritze and W. B. Robison, eds., *Historical Dictionary of Stuart England, 1603–1689* (Westport, CT: Greenwood Press, 1996).

33. Wouter J. Hanegraaff, *Dictionary of Gnosis and Western Esotericism*, Volumes 1 and 2 (Leiden: Brill, 2005), 199-201.

34. Michael E Hunter, Antonio Clericuzio, and Lawrence M. Principe, eds., *The Correspondence of Robert Boyle*, Vols. 1-6. (London: Pickering & Chatto, 2001), 454.

35. Oliver (1978).

36. Philip G. Rogers, *The Fifth Monarchy Men* (London: Oxford University Press, 1966).

37. Jacob (1998).

38. Yates (1979).

39. Arthur E. Waite, *The Holy Kabbalah* (New Hyde Park, NY: University Books, 1960). (Work originally published in 1902 as *The Doctrine and Literature of the Kabbalah* with Theosophical Society in London as publisher), 445 -452.

40. Waite, 447.

41. Yates, 21-22.

42. Z'ev Ben Shimon Halevi, *The Way of Kabbalah* (York Beach, ME: Samuel Weiser, 1976).

43. Perle Epstein, *Kabbalah: The Way of the Jewish Mystic* (Boston: Shambhala, 1988); Halevi (1976, 1997).

44. http://www.yashanet.com/studies/revstudy/rev6a.htm

45. Halevi, 25.

46. Yates (1979).

47. Halevi (1976).

48. Joscelyn Godwin, *Robert Fludd: Hermetic Philosopher and Surveyor of Two Worlds* (Boston: Shambhala, 1979).

49. Godwin, 34-35.

50. Yates, 23.

51. Matthew and Harrison, 575.

52. J.M. Batten, *John Dury: Advocate of Christian Reunion* (Chicago: University of Chicago Press, 1944).

53. Richard. H. Popkin, "Hartlib, Dury, and the Jews." In Mark Greengrass, Michale Leslie, and Timothy Raylor , eds., *Samuel Hartlib and Universal Reformation* (Cambridge: Cambridge University Press, 1994), 118 - 136.

54. A. Rupert Hall and Hall, *The Correspondence of Henry Oldenburg,* Vol. 1 (Madison: University of Wisconsin Press, 1965), 126.

55. Bernard S. Capp, *The Fifth Monarchy Men: A Study in Seventeenth-century English Millenarianism* (London: Faber and Faber, 1972), 23.

56. Philip G. Rogers, *The Fifth Monarchy Men* (London: Oxford University Press, 1966).

57. Archer published his conclusion in a book, *The Personall Reign of Christ on Earth* (1642).

58. Rogers, 13.

59. Rogers (1966).

60. Rogers (1966).

61. Rogers (1966).

62. Norman Penney, ed., *The Journal of George Fox* (Cambridge: Cambridge University Press, 1911).

63. Penney, 11.

64. Penney, 12-13.

65. Marjorie Hope Nicolson, ed. and Sarah Hutton, revised ed., *The Conway Letters: The Correspondence of Anne, Viscountess Conway, Henry More, and their Friends 1642-1684* (Oxford: Clarendon Press, 1912/1992), 161.

66. William Bray, ed., *Memoirs of John Evelyn,* Volume 3 (London: Henry Colburn, 1827), 280.

67. Rogers (1966).

68. Mede (1643).

69. Bray, 296-297.

70. Barbara Hannah, *Jung: His Life and Work* (Wilmette, Ill: Chiron, 1997), 315.

Chapter 8

Emerging Themes from the Work of the Women Alchemists

"Consequently, however astonishing it may seem, chaos seems to be essential not only for the survival of the organism, but for its development as well."[1]

I once wrote about how coming to the end of a research project was not really an ending at all. It is more like taking a snapshot of the work and thinking about what it can tell us at that time. The work is not concluded, just paused. Some themes emerged in the writing of this book about women alchemists that I would like to address as a means to conclude this part of their story—at least for now.

One More Look at the Scientific Revolution and the Mind-Matter Split

As we approach the conclusion of this book, I would like to revisit the Scientific Revolution briefly in order to place my thinking about the women alchemists in context. This is the time when the paradigmatic split between spirit and matter is identified as having its origin. One of my original questions regarding this time in history was how to reconcile that split within the context of the alchemical framework in which the natural philosophers thought about reality where *coniunctio* was the goal? This section could have been placed in the introductory chapters; however, when a reader is presented with new information and minimal context, the subtleties may get lost. The reader should now have more perspective on the issues that surrounded this pivotal age in history. We have spent some time with a few of the women alchemists and their stories; the following discussion will hopefully make more sense in light of the historical context of the 16th and 17th centuries. The preceding chapters have been a type of *separatio* operation where a large topic, alchemy, was teased apart into its diverse stories. Now it is time to bring these stories back together to see how they relate to each other, the *coniunctio*.

Edward Grant makes a strong argument regarding the connection between medieval science and the Scientific Revolution. A few science historians have tended to view modern science as a phenomenon that emerged on its own in the 17th century. Grant's point of view is significant because he shows that there was a flow of work between the two periods of time and that the assumption should not be made that the Scientific Revolution just happened on its own, independent of prior work. Whereas the early Middle Ages may seem to have been a shadowy time for intellectual growth, the time between 600 and 1200 CE was actually quite rich in debate regarding the nature of the universe and humanity's place in it as well as the exploration of one's relationship with God.[2]

We learned in chapter two that the advent of the medieval university (12th century) institutionalized a curriculum focused on the works of Aristotle that included natural science (also known as natural philosophy), the liberal arts, and theology. Grant makes a convincing case that the Scientific Revolution occurred because of the progress that was made in science in the late Middle Ages. This is also important in making a link between alchemy and modern thinking. The practice of alchemy was still very prevalent prior to and during the Scientific Revolution. Although its influence declined, alchemy still had a following that crossed generations to modern times.

Allen Debus writes an excellent description of scientific development during the period of the Scientific Revolution, which he dates as 1500-1800 CE, a time that includes, but is not limited, to the 17th century.[3] The Scientific Revolution was a time of rich discovery in the sciences. The Chinese had invented the magnetic compass (prior to 1040); it was improved and made popular in the West by the Venetians in about 1300.[4] The compass allowed sailors to navigate without the stars resulting in winter voyages under cloudy skies as well as making it possible for explorers to travel much further than they had been able to before. The world was opened up in a way that would not be seen again until the discovery of how to compute longitude accurately. Encyclopedias of both real and mythical animals were written and illustrated during this period. Botanical illustrations were also popular, often containing information useful to the physician as well as to alchemists preparing medicines. These volumes later included the observations of flora made by explorers and travelers. Samples of plants were brought home from far journeys and added to the growing pharmacopoeia. Thus, a classification system was needed for the thousands of plants being catalogued. Various methods were tried but it was Carolus Linnaeus (1707-1778) who finally created the system still in use today. Debus notes that despite the explosion of scientific knowledge, astrology retained a significant place in botanical theory. The reader will recall that in order to create healing elixirs, the Zodiac indicated the peak time to pick the herb.

The science of studying the human body experienced similar development during the 16th and 17th centuries. Galen's theories were taught in the universities although his work was never found acceptable by followers of Paracelsus. Galen believed disease was caused by an imbalance in the internal humors of the body whereas Paracelsus' teaching was that ill health was caused by external sources. Furthermore, the followers of Paracelsus criticized Galen who had made use of animals other than humans for dissection and thus made errors in his conclusions about human anatomy. Public dissections became popular in medieval times and were a requirement for medical students.[5] Galen's notion of the circulatory system was quite flawed and was challenged by William Harvey who published his famous work on the circulation of the blood, *De Motu Cordis,* in 1628. Harvey's work was not embraced universally and stimulated debate and further study, including some by members of the Royal Society of London. One proponent of Harvey's circulatory system was Robert Fludd (1574-1637), another well-known alchemist.

> Long interested in an aerial life spirit and its assimilation in man's body, Fludd had described a mystical circulation of the arterial blood as a   necessary consequence of the macrocosm-microcosm analogy in 1623 . . . but for Fludd it is evident that Harvey's anatomical evidence simply confirmed deeper mystical truths.[6]

Debus notes that Rene Descartes had also concluded that the heart was involved in blood circulation although his specific theory proved to be inaccurate. The main point is that in order for scientific understanding to move forward, older theorists had to be challenged, a process that continued with each generation and is still at work today.

The accomplishments of the 16th and 17th centuries would not be complete without mentioning the work of Nicolas Copernicus, Galileo Galilei, and Johannes Kepler. Their ideas shook the foundation of our notion of the universe by shifting paradigms from an Earth-centered universe to one whose center was the sun, heliocentric. The earth-centered universe versus one that was sun-centered would become a point of heated argument as well as in some cases, physical torture.[7] Aristotle had developed a comprehensive, if inaccurate, theory for the structure of the world as well as the motion of bodies. The Aristotelian system had first been revised by Claudius Ptolemy (c. 200 CE) who published *Almagest.* Although Ptolemy's system was more modern relatively speaking, he still placed the Earth at the center of the universe. The belief was that there could be no alteration in the celestial realm, that is, any position beyond the moon. Such thinking would become a flash point for disagreement between the Catholic Church and scientists such as Copernicus and Galileo. It was for Copernicus to reorient our movement to travel around the sun. He was followed by Tycho Brahe. Brahe made observations about a supernova that appeared in 1572 that placed it beyond the moon and in an area of the universe that was thought to be unchangeable, a puzzling dilemma. How could a supposedly unchangeable sector of the universe develop what appeared to be a new star? Brahe showed the planets orbiting the sun, yet the sun and moon still orbited Earth.

Johannes Kepler (1571-1630) refined the work of Copernicus and Brahe, calculating the planets' orbits mathematically and thus placing the sun at the center of our system. Kepler also had a mystical inclination and studied "musical harmonies and their relationship to the universe."[8] This was an area of natural science that linked music and the motion of planets. Finally, Galileo's mathematics supported Copernicus. His dealings with the Holy Office of the Inquisition are well documented, but suffice it to say, Galileo's work paved the way for Isaac Newton.

That alchemy was the paradigm for thinking about the universe throughout the emergence of the new science of the Scientific Revolution cannot be over-emphasized. Whereas alchemy was written about with healthy skepticism by some natural philosophers, the criticism was not so much at a theoretical level but very few people had been able to convince the populace that they had achieved transmutation of metals such as lead, into gold. In the *Memoirs of John Evelyn*, we find references to that argument. The reader will remember that Evelyn was a member of the Royal Society and contemporary of Robert Boyle. On December 14, 1650, Evelyn writes about having visited a Mr. Ratcliffe who had claimed to multiply gold but Evelyn "found him to be an egregious cheate."[9] On the other hand, Evelyn also relates a story told to him concerning a Genoese jeweler who could multiply gold. He adds that a goldsmith in Amsterdam claimed that:

> A person of very low stature came in and desir'd the goldsmith to melt him a pound of lead, which don[e] he unscrew'd the pummel of his sword, and taking out of a little box a small quantity of powder, casting it into the crucible, pour'd an ingot out, which when cold he tooke up, saying, "Sir, you will be paid for you[r] lead in the crucible," and so went out immediately. When he was gon[e] the goldsmith found 4 ounces of good gold in it, but could never set eye againe on the little man, tho' he sought all the citty for him. Antonio [Evelyn's source] asserted this with greate obtestation, nor know I what to think of it, there are so many impostors and people who love to tell strange stories, as this artist did.[10]

This excerpt illustrates how the idea of transmutation was not far-fetched in the minds of educated people of the day. Even a scientific thinker such as Evelyn was skeptical but neither was he dismissive of the possibility of transforming lead into gold. In a later entry from June 1705, Evelyn recalls his visit to Dr. Edmund Dickinson who was a well-known chemist at Merton College, Oxford. They discussed alchemy and Dickson claimed to have witnessed transmutation accomplished by a fellow who went by the name of Mundanus.[11] Again, their skepticism is not whether transmutation can take place; the discussion focuses on the success of the person doing the work.

Regarding the spirit-matter split in alchemical circles, Stanley Hall makes a connection between Robert Boyle, Christopher Wren, and Robert Moray as Freemasons.[12] He also claims that Moray belonged to the Rosicrucian Fraternity. Although lacking credible evidence, Hall argues that competing ideas about the theories of Francis Bacon resulted in a fracture between practitioners of alchemy. He claims the Rosicrucian Society moved into a more science focused orientation whereas the Freemasons leaned toward studies of the occult and esoteric works. In another reference to the matter-spirit split, Frans Wittemans (1938) claims that J.B. van Helmont, one of the well-known alchemists, tried to unite the "mystical Rosicrucians and the naturalist Rosicrucians."[13] I should add that Arthur Waite argues that "there is no traceable connection between Masonry and Rosicrucianism."[14] Waite claims that the Rosicrucians chose a scientific focus as well as an interest in magic. The Freemasons chose another focus not spelled out. These morsels of information are interesting in that they refer to a spiritual-physical split in groups linked to alchemical practices.

Another interesting story underscores how ubiquitous alchemical thinking was at the time, concerning Sir Thomas Gresham (1518-1579), founder of the Gresham College (1575) that developed into the Royal Society (he also founded the Royal Exchange). In 1543 he was elected a Freeman of Mercers' Company (still in existence). Gresham married Francis Bacon's aunt in 1544. Apparently, there was a time when the English Crown was in great need of funds due to the costs of war, poor management, and inflation. The merchants were wealthy and acted as lenders to the Crown. Gresham, as Royal Agent, proposed that the Crown should obtain a monopoly of lead.[15] Why lead? Gresham provided free lectures to the public. One of the lecturers was John Dee, the famous explorer in alchemy and the occult. The Crown did not achieve a monopoly in lead; yet, one wonders whether there is a connection with alchemy and the belief in transmutation. Could this be linked to an alchemical dream of turning lead into gold? This story is just one example of the kinds of intriguing vignettes that emerged that illustrate the thinking of those times in history. These ideas were not viewed as preposterous in the way they would be seen today.

The Women Alchemists' Path Toward *Separatio*

Regarding at least the women alchemists of the seventeenth century, Lynette Hunter, concludes:

> Another interesting factor linking all of these women [Dorothy Moore, Anna Maria von Schurman] is their early relation to Descartes' work on reason, mind and the body, which each writer later abandons. Moore's own development indicates a tense recognition of the separation of mind and body as a symptom of spiritual illness and political corruption, with a use of it as an analogy for the different actions of spirit and body in the workings of grace.[16]

Historians often think of the Scientific Revolution as the time when the spirit-matter split attributed to Descartes, gained strength and continued to be the paradigm for scientific thinking into modern times. Matter would come to be seen as behaving like a machine; spirit and mind would be something else en-

tirely. I have wondered how natural philosophers who worked within an alchemical paradigm would fall into that line of thinking. Alchemy is about separating spirit and matter but the work is not complete until all the parts are reunited. The reader is now familiar with *separatio*, the operation in which the
alchemical matter undergoes distillation in order to separate the pure from the
impure matter. In other words, "order is brought out of confusion"[17] or what one
may perceive as chaos. F. David Peat explains that in chaos theory, the chaotic
state is not random and meaningless. Chaos is actually a highly complex state
that only looks random until one understands its underlying patterns.[18] John R.
Van Eenwyk also illustrates how seeming chaos actually precedes a state of
greater complexity and stability.[19] He suggests that one of the outcomes of the
shadow in psyche is a jangling of our thinking, creation of a little chaos, which
in the restructuring and coming to terms with our complexes creates a new and
different element in our consciousness.

Van Eenwyk's explanation is very much like the way the alchemists described the result of *coniunctio* where a third element was created after the steps
of separation and re-combination of the substances being worked upon. I believe
that by looking at the split in a more holistic way, we can surmise that the
women alchemists, as well as the field of alchemy, had to undergo a phase of
separatio before the disparate parts could be re-integrated. What does that
mean?

The naming act of Logos is a *separatio* event. When one names something,
it becomes one thing and not another. For example, if one describes an event as
raucous, the possibility that it might have been calm must be dismissed. Depth
psychology teaches that psychologically, *separatio* occurs with the evolution of
consciousness of the opposite contents of the psyche such as the self-awareness
of one's light and shadow qualities.[20] In order for psychic development to evolve
in a healthy direction, we need to become aware of, and understand, negative
qualities that we may not like about ourselves. As well, positive qualities that
we see in others but have not accepted in ourselves must be integrated into our
self-concept. *Separatio* is often symbolized in the old alchemical engravings by
a sword or by contrasting images such as sun and moon or king and queen. Edward Edinger explains that concrete and symbolic meanings of action can reflect
separatio. For example, physical alchemy was concrete, whereas spiritual alchemy was symbolic. In time, these two aspects of alchemy began to seem in
such opposition to each other that practitioners chose to practice one or the
other, not both; they were split or separated. Thus, we see the emergence of
modern science and specifically, chemistry, that has struggled with anything that
is not measurable, able to be manipulated, or is concrete, as well as the lesser
acknowledged esoteric spiritual alchemy that is practiced, for example, by modern Rosicrucians.[21]

Metaphorically, one must be able to separate qualities such as light from
heavy in order to understand their opposite, yet interconnected, relationship.
Jung writes:

In the beginning God created one world (unus mundus). This he divided into two—heaven and earth . . . . The division into two was necessary in order to bring the 'one' world out of the state of potentiality into reality. Reality consists of a multiplicity of things. But one is not a number; the first number is two, and with it multiplicity and reality begin.[22]

Edinger explains that the more formal description of opposites emerged from the Pythagoreans and that these qualities had a numinous property. Some of the opposite qualities they spoke of were "limited/unlimited, odd/even, one/many, right/left, male/female, resting/moving, straight/curved, light/dark, good/bad, square/oblong."[23] Edinger also employs a helpful analogy comparing the energy of a battery's two poles to the psychic energy of the opposites. Energy exists by means of movement between the two poles.

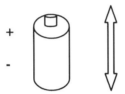

Figure 8.1: Diagram Illustrating Edinger's Battery Analogy

The significance of recognizing that opposite qualities also exist in one's psyche cannot be underestimated. Edinger adds, "The world has to be rent asunder and the opposites must be separated, in order to create space in which the human conscious ego can exist."[24] Consequently, "the young ego is obliged to establish itself as something definite and therefore it must say 'I am this and I am not that.'"[25] The *not* qualities may become part of the shadow which will eventually need to be confronted during the individuation journey when the psyche is more able to tolerate that flow of energy between the ego and the unconscious, not to mention the great collective or world unconscious.[26] *Separatio* is the opposite of *coniunctio*. The holding of the opposites is a *coniunctio* process, similar to the way the battery displays opposite polarities while remaining connected as one unit. Jung believed that developing the ability to contain that dance of the opposites in our psyches was crucial for the creation of consciousness and was the task that modern humanity must accomplish.

James Hillman also speaks to the *separatio* operation when he reminds us of the multiplicity of the unconscious, its myriad archetypal images.

By employing the dream as model of psychic actuality, and by conceiving a theory of personality based upon the dream, we are imagining the psyche's basic structure to be an inscape of personified images. The full consequences of this structure imply that the psyche presents its own imaginal dimensions, oper-

ates freely without words, and is constituted of multiple personalities. We can
describe the psyche as a polycentric realm of nonverbal, nonspatial images.[27]

The recognition that our psyche contains multiple sides is not necessarily pleasant when it first occurs. Depth psychological psychotherapy asks the analysand
to come into relationship with those aspects such as the *inner child* one often
hears referenced in popular culture. Hillman describes the disorienting experience "of the multiple persons of the psyche."[28] One then builds relationships
with these archetypal images. This may be the operation that is currently being
played out in modern science, a rebuilding of the relationship between spirit and
matter.

Hillman's exploration of the meaning of soul and spirit also reflects a *separatio* image. He sees the soul as depth, descent, feminine, and water whereas the
spirit is of the ascent, light, masculine, and fire. Hillman characterizes the spirit
as a mountain peak whereas he perceives the soul to reside in the vale or the
valley. The spirit soars; the soul is in deep reflection.

The *separatio* that occurs within the context of the scientific method is also
a discernment. It was inevitable that as the scientific method required the observer to objectify his subject, that a split between that which can be observed
(matter) and that which cannot (soul/spirit/mind) would occur. It was necessary
for a mind-body split to occur in order for each to be more fully understood.
Veronica Goodchild notes that the split allowed great advances as well as the
concomitant challenges in both the material world and later, psyche. She writes:

> We live in an age that has produced the demonic outpourings of the radical split
> between spirit and matter. Like two galaxies hurtling away from each other
> over the last two thousand years, spirit and matter, having forgotten their ori-
> gins in a unified whole, have degenerated into rampant materialism and rational
> intellectualism. Although this separation allowed an extraordinary acceleration
> of knowledge, and the differentiation of an ego consciousness disentangled
> from the things of nature and the world, the cost to the soul is inestimable. We
> need only to mention the release of the atom bomb, the extermination of mil-
> lions of people in two world wars, and the continuing atrocities that mark our
> present times. We live in an age of breakdown.[29]

Goodchild's concern echoes that expressed by many of the women alchemists. What should be apparent is that like all alchemical operations, *separatio*
is *necessary* for the work to be successful. In order to be able to understand the
observed, one *must* be separated from it. The problem arises when one becomes
stuck in the *separatio*, losing the sense of connection that exists. That connection will be further explored shortly regarding the *coniunctio*.

The point is, just as in alchemy where each operation must be completed in
order to achieve ultimate success, the *separatio* that emerged between spiritual
and physical alchemy was necessary. The path to *separatio* had to be traversed
because that was the next step. In order for Western scientific development to

proceed as it did, old ways of thinking had to be distilled and analyzed. A new paradigm can emerge once the elements are reunited. Spirit and matter needed to be explored separately before the connections could be perceived and understood. The women alchemists were probably unconscious of the way their work bridged the divide between spirit and matter with its focus on healing, theology, alchemy, Kabbalah, and Apocalyptic study—the intertwining of spirit and matter.

Mary Anne Atwood: In Between

"Alchemy is a philosophy; it is the philosophy, the seeking out
of THE SOPHIA in the mind."[30]

The story of Mary Anne Atwood provides a final illustration of the struggle engendered by *separatio* as physical alchemy became established as the field of chemistry. Spiritual alchemy was destined to be linked to occult studies, which eventually were concluded to be mere superstition. Once these two facets of alchemy became separated, with one being marginalized whereas the other was regarded as hard science, the two facets became fractured and have yet to be reunited.

Mary Anne Atwood (née South) was a 19th century woman alchemist who found herself caught between the two worlds of alchemical philosophy.[31] Mary Anne's father, Mr. Thomas South of Gosport, Hampshire studied religion, philosophy, and alchemy. South believed in *prisca sapientia*, or that the ancient philosophers held the secrets that were passed on to the alchemists. Mary Anne studied alongside her father and also became convinced that the ancient thinkers learned some secret or truth regarding humanity's connection to the Divine that they passed down to those who could understand. They tended to lean towards a more spiritual interpretation of the work, believing that the true goal of alchemy was for imperfect humanity to be made perfect. Yet, she also studied physical alchemy comprehensively. She published *A Suggestive Enquiry into the Hermetic Mystery* in 1850 in London at the age of 37 years. Subsequently, she and her father decided to either ask for the return of her book or bought back as many copies as they could locate and burned them as well as a poem he had written on alchemy. Arthur Burland suggests this was a result of, "her fears that she had stumbled on a great secret and had thoughtlessly revealed too much. She refused to speak more on the subject of her studies, even though she lived until 1910 when she died at the age of ninety-seven."[32] It is not clear what Mary Anne and her father feared specifically but they had concluded that their work held truth that was beyond that which the current generation could tolerate. Burland links Mary Anne to Jung, claiming that she:

Realized that the real aim of the alchemists was to induce a state of enlightenment in which they became at one with the universe and could be reached di-

rectly by God. She also understood what the adepts meant in their many warnings that the path of enlightenment was inward.[33]

Perhaps, Mary Anne represents the slide of the science of alchemy toward its
more modern occult status. Nevertheless, *Suggestive Inquiry* did survive and
was reissued in Belfast (1918) with a third edition in 1920. The 1920 edition
contains a lovely introduction by Walter Leslie Wilmhurst who included the
quote at the beginning of this section in his explication of Mary Anne's work.

The *Suggestive Inquiry* is a comprehensive discourse on myriad facets of
alchemy. It begins with the history of alchemy and its genesis in the East making connections to early Egyptians, King Solomon, the Kabbalah and the Eleus-
inian mysteries. Mary Anne quotes *The Emerald Tablet* of Hermes Trismegistus
and even touches upon the work of Maria the Jewess, referring to her with one
of Maria's many monikers, "*Maria Practica.*"[34] Mary Anne's scholarship is
wide-ranging; Roger Bacon, Basil Valentine, Raymond Lully, Albertus Magnus,
Nicholas and Pernelle Flamel, Agrippa, and Spinoza are each commented upon.
She touches on the bogus practitioners of alchemy as well as those who believed
that transmutation could lead to wealth. Mary Anne reports the stories about
transmutation clearly and objectively. It would appear that she believed it was
possible and hesitated to outright deny the possibility in light of the first-hand
accounts of transmutation that she quotes. However, she also acknowledges that
there she is not aware of any physical evidence to be able to conclude it was
indeed possible to turn lead into gold using the Philosopher's Stone.

The bulk of the *Suggestive Inquiry* explains Mary Anne's spiritual philosophy regarding the true nature of alchemy. It is beyond the scope of this book to
present her thinking with the depth that it deserves. Yet, she includes a random
thinking point in her appendix that she titles, "Table Talk and Memorabilia of
Mary Anne Atwood, Begun August, 1860."[35] Note number 20 summarizes her
beliefs rather succinctly.

> We must remember that Alchemy is Divine Chemistry, and the transmutation
> of Life; and therefore that which is the medium between soul and body is
> changed, and the soul freed from the chains of corporeity, and the body is left
> as a mere husk. These people put on their bodies as mere coats.[36]

To summarize this section, some depth psychologists as well as many spiritualists lament the mind-matter split. I have often heard Isaac Newton referred to
as a brilliant and rational scientist whose work in the occult was misguided or
something to be ignored. He is also often blamed for the spirit-matter split.
Newton had straddled the line between the new science and ancient beliefs about
the nature of the universe. His more occult work brings to mind the story of a
modern physicist, F. David Peat.

In *Blackfoot Physics*, Peat relates the journey that he took in his understanding of the science of Native Americans or more accurately, the First People.[37] He discovered that indigenous natural philosophy had much in common with modern conceptions of the universe held by theoretical physicists. Perhaps *coniunctio* in our thinking about reality is closer than we might think, always keeping in mind that there are unknown realities still to be discovered. *Coniunctio* will be the final operation we address in the next section.

Closing the Circle: *Coniunctio*, Mystic Sisters and *The Chrysopeia of Mary the Jewess*

The uroborous is often portrayed in alchemical illustrations as a dragon or serpent eating its tail. The final stage of alchemy is the *coniunctio* which is also a closing of a circle of operations that began with dissolution. This concluding section focuses on the final operation.

Clark Moustakas and his work in heuristics has been like a friendly voice in my head while doing this investigation and he reminds us that the research process is incomplete until we talk about how the work has affected the researcher personally. This is something I ask my Master's degree candidates to do at the end of their thesis—talk about how they have changed by doing the work. I explained in the Preface how writing this book had presented some major trials for me, mostly of my own making. In this section I wish to relate some of the story of the cover for this book. It addresses so many of the elements I have been writing about including one's personal journey and is a lovely example of synchronicity. Finally, the reader might hear something useful in the telling of how this work has affected me.

Nearing the completion of this book, a remarkable Jungian analyst directed me toward a painting of Maria Prophitissa/Maria the Jewess, the historic Jewish woman alchemist described in chapter three. The painting, *The Chrysopeia of Mary the Jewess* was painted by Leonora Carrington in 1964. Her work was recently featured in an enormously popular showing of Surrealist artists at the Los Angeles County Museum of Modern Art (LACMA) titled: *In Wonderland: The Surrealist Adventures of Women Artists in Mexico and the United States*, January 29–May 6, 2012.[38] I was unable to attend the exhibition but when I observed an image of her painting I was completely mesmerized. Carrington had brought together all the elements I had been writing about concerning the women alchemists: alchemy, the Kabbalah, Apocalyptic images, science, and feminism. The images are fantastic and there were many I was unsure about. However, I felt strongly that *The Chrysopeia* would be so wonderful for the cover of this book about women alchemists and their journeys.

I was guided by one of my sisters who is a much more effective Internet researcher than I, toward an analysis of *The Chrysopeia* written by Dr. Gloria

Orenstein. I contacted Professor Orenstein to find out if she knew who owned *The Chrysopeia*. It turned out that she was a dear friend of Leonora Carrington's for over thirty years (and has an amazing life story herself). Gloria Orenstein is a distinguished professor of Comparative Literature at the University of Southern California and an expert on women Surrealist painters. I could not believe my luck! She was kind enough to direct me to the Curator for the exhibition, Dr. Ilene Susan Fort, who also wrote and edited the book, *In Wonderland: The Surrealist Adventures of Women Artists in Mexico and the United States* (2012).[39] Dr. Fort facilitated my being able to contact the painting's owner who graciously gave me permission to use an image of *The Chrysopeia* for the book cover. Each of these women now holds a part of this work. Moreover, in talking with them, I was astonished to learn of a whole world of women Surrealist painters who have gone unacknowledged, something that resonated deeply with the work of the women alchemists who mostly went unrecognized and undocumented. The connection between the painting and this book was quite meaningful to me. In order to better understand the artist, I began exploring Leonora Carrington's life to discover how she came to paint such an extraordinary depiction of the nexus of alchemy, feminism, and Jewish mysticism. Her work, as well as that of her friend and brilliant Surrealist, Remedios Varo, embodied my understanding of the journeys of the women alchemists. Moreover, *The Chrysopeia of Mary the Jewess* closed a circle in my own work. Maria Hebraea appeared in the beginning of the women alchemists' stories and *The Chrysopeia* had now appeared at the end of my work. Each step of the process seemed a bit magical and depth psychologists will recognize synchronicity at work.

Mystic Sisters

The three women discussed in this section are, and were, modern alchemists in their own way. Leonora Carrington was born in Clayten Green, Lancashire near Chorley Lanes, April 6, 1917.[40] She studied art in 1936 with Amédeé Ozenfant. Leonora met and fell in love with Max Ernst in Paris 1937, which is how she was introduced to Surrealism. Ernst was interred as an enemy alien when WWII broke out. Leonora escaped to Spain but suffered a nervous breakdown and lived in a private clinic. She wrote a book about her terrible experiences titled *Down Below* (1944). Leonora was married in 1941 to Renato Leduc, a Mexican poet, in what is described as a "marriage of convenience."[41] They divorced in 1942. Leonora lived for a while in New York but soon traveled to Mexico where she spent the rest of her life. She became close friends with another wonderful Surrealist, Remedios Varo. Leonora married a Hungarian, Imre Weisz, a photographer and Holocaust survivor in 1946.

Carrington wrote prolifically and painted a remarkable body of work. Orenstein reminds the viewer to pay careful attention when analyzing Carrington's art as she included alchemical symbols, esoteric images, Gnosticism, magic, runes, and imagery from many spiritual traditions in her work. Carrington brings these disparate images into *coniunctio* as they swirl around each other on the canvas. I will discuss *The Chrysopeia of Mary the Jewess* shortly.

Remedios Varo (1908-1966) was another amazing Surrealist painter who like Leonora, tackled alchemical symbolism in her work. She was born in Girona, Spain but lived in Mexico City until her death.[42] Varo began her artistic training in 1934 at the Reál Académia de San Fernando in Madrid. She crossed paths with the Surrealists living in Paris and in 1937 married Benjamin Péret, a poet. In a manner also quite similar to Leonora's experience, as a result of Germany occupying France in 1942, Varo and Péret escaped to Mexico to a colony of Surrealists in exile including Leonora Carrington. By 1953 Varo began painting full time. Jane Turner writes that Varo was "influenced by André Breton in her cultivation of dream-like moods, but she rejected an unswerving reliance on the subconscious in favour of deliberate fantasies."[43] Varo's work is astonishing. For example, *Creation of the Birds* (1957) illustrates her infusion of alchemy, the occult, nature, and the strength of the feminine in a truly stunning way. One of my favorite images in the painting shows a distillation apparatus that creates paint by distilling some type of matter from outside the painter's studio window. She is depicted as part woman-part owl. As the owl-woman paints birds, she shines a magnifying glass in the shape of the alchemical symbol for sulphur that focuses light from either the moon or stars. The light brings the birds she is painting to life.

Gloria Feman Orenstein, the third mystic sister (or *soror mystica*) in this section, met Leonora Carrington in a rather unusual way and I direct the reader to her article, "The Many Worlds Of Leonora Carrington: Navigating Shamanic Journeys and Surrealist Border Crossings" which was a paper given at the ACLA conference in Puebla, Mexico. It is a wonderful story. Suffice it to say, after many attempts to meet Leonora, the artist turned up in New York City and called Gloria Orenstein saying, "This is Leonora Carrington. I have just arrived from Mexico and would like to meet you." [44] There began a 30 year friendship that I hope will be recorded someday as it is full of rich imagery and profound depth of thinking.

I have mostly focused on women alchemists prior to the 18th century in this book. These three women are lovely examples of how the sisterhood of the *Soror Mystica* is still alive, albeit transformed.

The Chrysopeia of Mary the Jewess

Carrington's work shows elements of feminism, esotericism, study of the Kabbalah, a great emphasis on alchemy, and mythology or as Orenstein comments:

> I often refer to her oeuvre as a kind of CODEX OF FEMINIST AWAKENING. In saying that I refer not only to accessing the powers of deities such as the Goddess, and the animal powers, but also to the way in which art, itself, becomes, as she has said 'her vehicle of transit.'[45]

I would like to direct the reader to the cover of this book to take in the imagery that is Carrington's gift. I felt a kindred spirit when I read Orenstein's statement regarding *The Chrysopeia*; "This is obviously a very rare artistic depiction of a Jewish female Alchemist in all of western art history."[46] Orenstein relates that Leonora Carrington was very interested in the idea that alchemy transmuted elements in the way that cooking and eating transformed food. Eating and digestion were actually alluded to by early alchemists when they were describing transmutation. As well, Leonora's second husband was a Shoah survivor and her two sons identify as Jewish. There is considerable reference to these ideas in the painting. There is so much to see in *The Chrysopeia* and Orenstein stated that Carrington never explained explicitly what it meant. There are alchemical symbols, the great oven or athanor, and images from the Kabbalah.

Orenstein observes some remarkable things about the painting. She points out that Mary the Jewess is painted as half woman-half lion, depicting the merging of the masculine and feminine. Carrington believed that much of humanity had lost its connection to its animalistic self which she believed was resulting in the destruction of countless species and having such a negative impact on the planet. Mary the Jewess maintains that connection with her planet that is a living thing. Moreover, Mary is seemingly surrounded by the Zodiac, a pentagram and a mandala-like circle that might allow her to call on multiple traditions in her work. She wears a collar whose ring for a chain is still intact. There is no chain but she must do this work. Light and energy seem to emanate from her hand.

Another fascinating image is at the top of the painting. One can see a circle transcribed in what might be a hexagon. I encountered a similar diagram to this by Franz Dornseiff in his 1922 book, *Das Alphabet in Mystik und Magie*. Inscribed in the circle of his diagram are a square, triangle, and a hexagon whose intersections with the circle denote the 12 signs of the Zodiac.[47] I cannot claim that Carrington saw this diagram; however, the connection to astrology is consistent with her studies. In the same image of the painting, there are two hands emanating from the circle that also seems to include the eyes of some feathered being. In older alchemical engravings, one often sees the "hands of God" bestowing a blessing on the work or possibly providing needed Divine assistance. Carrington's hands seem more feminine with the lacy patterns and feathers. Might she have been depicting the "hand of the Goddess"?

I am also captivated by the giant creature to the right of Mary who is standing over both her and three other bearded figures. It appears that there is a downward pointing triangle with a dot in the middle on his chest which is similar to the alchemical symbol for blood. He also sports the infinity symbol above his eyes. There are numerous details one can speculate about for hours.

Orenstein writes about the Nativity aspects of the painting, as well.

> The scene in the Chrysopeia can also be viewed as a Nativity symbol with the three wise men in attendance, the inclusion of baby animals, and the birth of the Philosophical Child. These wise onlookers seem riveted to the spot as they ob-

serve the Alchemical performance of Mary. This Mary, the Alchemist, a Jewish woman, has taken her power back from the males, who, in traditional Judaism are usually the ones empowered to perform sacred rituals. Here she is the Woman of Power who performs the sacred ritual instead of the men.[48]

Carrington does not romanticize alchemy. Just as there existed malicious practitioners in former times who used alchemy for their own ends, usually financial gain, Orenstein believes that the athanor in *The Chrysopeia* carries the shadow side of alchemy. Near the rear of the scene there is an insect-like, spaceship type object that is an athanor (oven). It has the alchemical symbol for salt inscribed near the top. There are flames in the four windows/openings in the main body with what looks like a door at the entrance. Two robed figures are standing next to another figure lying on some type of table, covered in bandages. It appears to be a woman and two hounds, one seeming to bay, are her escorts.

Orenstein sees the athanor as a crematorium. In this instance, Hitler used the oven to destroy a people as well as obtain gold in a horrific act. I tend to agree with that analysis due to further examination of more symbols present in the painting. I have used the early alchemical notations discussed in chapter five to try to interpret the work. Although they are difficult to pick out, there is an inscription painted above the windows. The symbols seem to come from a variety of systems including alchemy, Greek, and astrology. Indeed, there is a crab that sits between the uppermost windows that is akin to the astrological sign, cancer. Jean Chevalier and Alain Gheerbrant write about the symbolism of the crab which is associated with the moon in most cultures and as well, the feminine. They also discuss the Zodiac house of Cancer stating:

> Cancer also plays the part of intermediary, marking as it does the middle of the year, linking what is to what will be, the threshold of reincarnation, the way from zenith to nadir. Those influenced by this sign enjoy strong and hidden powers potentially favourable to future incarnations.[49]

The alchemical operation of calcinatio is a purification of the substance being worked upon that will undergo further transformation. However, the success lies in the Adept working with the Divine and not for his own gain.

Orenstein argues that Mary the Jewess is performing a cleansing ritual that will also protect the Philosophical Child to be born. She suggests that the salt symbol on the crematorium is there to invoke needed purification. The mercury symbol on the tall structure to the left and the sulphur symbol on the cage where one creature is trapped while another seems to try to escape, make up the three needed ingredients to do the work. The reader will remember that alchemists represented the soul, body, and spirit of matter with these three symbols as well. They believed all matter was made of philosophical salt/body, the sulphur/soul, and mercury/spirit.

In our attempt to hear the message of *The Chrysopeia*, one must also study it with the Kabbalah in mind. Orenstein states, "she [Carrington] has always told me [the Kabbalah] was fundamental to understanding many of her paintings."[50]

> If Mary the Jewess had a knowledge of Cabbala similar to that of Leonora, she would have known that the task of the Tsaddik, or the righteous person, similar to that of the saint in Christianity, is to repair the world by restoring the fallen sparks of light to the Godhead, or by enabling the spirit of God to be united with the Shekinah, the feminine aspect of God, that has been in exile. When the Shekinah, which is in exile, but which is conceived of by the Cabbala as the community of Israel and as the feminine aspect of God, is reunited with the rest of God (the other Sephiroth of The Tree of Life), then the Messianic era will commence. . . . THE CHRYSOPEIA OF MARY THE JEWESS (1964) clearly shows us that the work of this Jewish female Alchemist enables the lifting of Divine sparks by immobilizing those negative energies, entities and forces that would annihilate them. In the painting we see the divine sparks rise as Mary's Jewish alchemical and cabbalistic work sets the process in motion. The Divine sparks are the positive energies that enable the Shekinah to be reunited with God.[51]

One might see the Kabbalistic Tree of Life in the form of the oven, as well. Its chimney reaches up to the heavens, in the place of the Sephirot called Keter, the first emanation from God. The door is in the position of Malkhut where the feminine aspect of God resides, the Shekinah. They are joined in this image. Mary's work must take place for the world to be healed. Therein lays the *coniunctio* in that the female aspect of God must be reunited with the masculine if balance and harmony are to exist. That is the *coniunctio* that is so evident in this painting.

I found Carrington's work, as well as that of Remedios Varo, to contain images of the women alchemist's journey in that the strength of the feminine going back to antiquity is portrayed so beautifully. Carrington's work illustrates the different themes of this book: alchemical symbolism, religious beliefs, and one's journey to come into relationship with the Divine. Her works make me think of the loss that occurs when rational thinking rejects any point of view that seems irrational. Both the feminine and the masculine in psyche lose when one or the other is cast aside.

Conclusion: *Finis Coronat Opus* (The End Crowns the Work - Ovid)

Edinger reminds us that "the *coniunctio* is the culmination of the opus."[52] Physically, the *coniunctio* resulted in the creation of the Philosopher's Stone. Psychologically, the *coniunctio* is about the creation of consciousness. Jung describes three stages of the *coniunctio*. First the spirit and soul are united. This is followed by spirit and soul further uniting with the body, "but a consummation of the *mysterium coniunctionis* can be expected only when the unity of spirit, soul, and body is made one with the original *unus mundus*,"[53] the third stage. The *unus mundus* refers to the interconnectedness of all things. It is the primal or chaotic state that is the beginning and end in Kabbalah and alchemy. Material and spiritual phenomena do not exist in separate states.[54] In that place we find the phenomenon of synchronicity. Jung discusses synchronicity or "meaningful coincidence"[55] in the context of the *coniunctio*. "Everything that happens, however, happens in the same 'one world' and is a part of it."[56] The world as *unus mundus* is connected in ways modern physics is still trying to understand.

Edinger describes a lesser *coniunctio* and a greater *coniunctio*. The lesser *coniunctio* is a preliminary "union or fusion of substances that are not yet thoroughly separated or discriminated."[57] Identifying the shadow parts of the psyche is an example of a phase of psychological work that requires further effort, a lesser *coniunctio*. "The lesser *coniunctio* occurs whenever the ego identifies with contents emerging from the unconscious."[58] This can be demonstrated by over-identification with specific people or groups where individuals may no longer be able to think of themselves as existing apart from some philosophy or fanatical ideology.

Further psychological death and separation must occur in order to develop a healthy psyche. A *mortificatio* phase occurs as more separation must take place and individuals begin to be able to be more conscious of their thinking and actions. The greater *coniunctio* follows this final separation phase and the work becomes complete. Rather than thinking of psychic development as linear, it is important to recognize the cyclical nature of the work. A psychological complex may have to be revisited several times before it is resolved. Furthermore, various parts of the psyche may be at different stages developmentally.

Experiencing the opposite contents of the psyche, simultaneously, is characteristic of the greater *coniunctio*. It is often represented symbolically by the marriage of opposite qualities expressed as king and queen or lovers. The ego plays a critical role in this process. "The ego brings about the union of opposites and thereby creates the Self, or at least brings it into manifestation. Thus is underscored the supreme importance of the conscious ego."[59] Hillman describes what he calls the "puer-psyche marriage"[60] which is a union of spirit and soul. It is a recognition that soul and spirit need the qualities of each other.

> The spirit asks that the psyche help it, not break it or yoke it or put it away as a peculiarity or insanity. And it asks the analysts who act in psyche's name not to turn the soul against the puer adventure but rather to prepare the desire of both for each other.[61]

The place where the third or Philosophical Child exists is the domain of the imaginal.[62] It is a place between spirit and matter, not one or the other. Robert Romanyshyn contends that alchemy "is a kind of consciousness which holds this tension and in holding it the subtle body of the third, the soul, the realm of the imaginal, which is neither that of spirit, consciousness, mind, nor matter, nature, body is born."[63] He adds that the alchemists lived in this place of the imaginal and it is simple to see how this could be so. It does raise the question, however, regarding projection. Was alchemy all about projecting unconscious contents onto the work or was it more about spending time in the world of Jung's subtle bodies and the imaginal? Perhaps it is best to follow Romanyshyn's example by not falling into an either/or dichotomy but to consider that alchemy was about both phenomena. It certainly suggests that alchemy was far more than the precursor to modern chemistry which is how it is typically described today. Alchemy seems to be "in that same place of neither/nor, the third of metaphor."[64]

Patricia Taylor has written about the in-between place or *metaxy* in her work on the archetype of *Enough*.[65] She discusses how the act of dialogue creates a space that is ripe for the work of the unconscious. When people reach together to find meaning, the *metaxy* is there and the work is to try to stay in that in-between place, for at least a while.

Finally, Jung explores the difference between spirit and matter as if they are *separatio* images.[66] He states that he was struggling with the concepts and concedes that the terms can be defined approximately at best. Interestingly, Jung argues that mind and body should be viewed as "two sides of the same coin"[67] contrary to the notion that they exist as separate opposites. This makes sense and is consistent with alchemy in that the opposites come together and form the third, which is neither one nor the other. One of my original questions concerned how the mind-body split of the Scientific Revolution could occur within an alchemical context. Perhaps the early thinkers spoke of mind and body but thought of them as being connected in a way that modern readers have not understood. Somewhere in time, and all indicators point toward the 17th century, mind and body were no longer perceived as different sides of the same coin but became disconnected. This separation of mind and body is usually attributed to Rene Descartes. It was a necessary step yet, it is as if the world psyche became stalled in the *separatio* phase of alchemy. Nevertheless, we know that the next step in the process is the lesser *coniunctio*. Perhaps we are witnessing the emergence of this phase in modern physicists' recognition of the interconnectedness of matter and spirit as well as the work of Jung that resonates so well with many readers. This has important implications for future generations. For the sake of argument, if we accept that the world psyche is entering a lesser *coniunctio*, it is clear that further *mortificatio* and *separatio* must take place before a greater *coniunctio* can occur. Will our children and their future offspring be prepared to experience the dark times that must surely follow? What can the elders do to reassure the children that they are not alone in the alchemical vessel?

Carrington and Varo concretize in their art the nexus of the divine feminine, alchemical operations, and even their relationship with the masculine energies in the world. If alchemy teaches us anything, it is that both the feminine and masculine principles (not gender related) are necessary for the work. The repression of the feminine does not serve us well as it severs our connection with half of our psychic heritage. The Philosophical Child cannot be born from one or another but must have the two parents joined.

Recognizing the work of the women alchemists deepens our understanding of not only this old science but it completes the history of the development of scientific thinking. Virginia Woolf stated this beautifully.

> For there is a spot the size of a shilling at the back of the head which one can never see for oneself. It is one of the good offices that sex can discharge for sex—to describe that spot the size of a shilling at the back of the head.[68]

Did the women alchemists have significant impact on history? Perhaps not if one relies solely on traditional publications. The evidence is there that they did engage in scientific study and as well, shared their ideas with their male counterparts. Much has changed from when Virginia Woolf contemplated the status of women in her time. My aunt was active in women's issues and women like her and her generation are the giants upon whose shoulders we stand, just as Newton exclaimed regarding the thinkers who preceded him. The women in my book were relatively ordinary if one compares their accomplishments to that of their more famous brethren. Yet, they were very brave although they would probably say, not so. She who steps outside the mores of one's time crosses back and forth over the borderland. Creating that seemingly odd art, running the school that is so different from its traditional peers, writing the sentences that inflame, these are not easy things to do. But they must be done. The women alchemists knew this. The work was a vocation; the thoughts had to be thought. The only choice would have been to follow a path that led to the death of the soul. Wherever one's inner spark derives, trying to repress or extinguish it kills something precious and a gift from wherever one believes gifts come. How does one say to the universe—no thank you? Anyone who makes that mistake knows she has killed something and will carry it to her grave. The women in this book may have endured criticism, dismissal, and/or contempt but they said thank you for the gift and let it lead them in untold directions.

Exploring the stories of the women alchemists seemed chaotic as I began the task. I had so many questions: Why did they practice alchemy? What kept some of the women grounded in healing whereas others pursued more esoteric research? How did the women who concentrated on their questions about the nature of the universe feel about being mostly watchers from the sidelines? How did they deal with being kept out of the formal scientific societies and having only limited access to publishing?

I would like to think back to the questions I first posed about the *soror mystica* and synthesize what I found. First, there were many women throughout human history who sought to understand the nature of the universe as well as humanity's place in it. These women came from all walks of life but some were more fortunate and had the means to learn to read and carry out their work. Finding evidence of their work in non-traditional places has required getting to know them more personally. Their work could not be separated from their families, lovers, politics, and their theology. Attempting to understand why each woman studied her art involved listening to their stories in a deep and profound way.

I also learned that the link between theology, science, and politics was every bit as intertwined and convoluted as it is in this century. It is time consuming but to really understand one element completely, all three elements must be looked at individually before making sense of them as a whole.

It is also clear that I have only begun my own work. There are many stories to be learned from other times and other cultures. For instance, as I mentioned in chapter one, there is evidence that women practiced alchemy in China such as Thai Hsuan Nu, Fang (1st century BCE), Keng Hsien-Seng (975 CE), Pao Ku Ko (c. 200 CE), Li Shao Yan (c. 1000 CE), Thaittsuan Nu, and Sun Pu-Erh (c. 1100 CE).[69] Fang came from a family that was well versed in alchemy and she studied with a wife of the Emperor Han Wu Ti. Fang seemed to be adept at distilling silver from ore using mercury, a common alchemical practice that appeared to create silver from mercury. It is said that Fang's husband became obsessed with learning his wife's secrets to the point of physical torture. She finally went insane, possibly from mercury poisoning, and killed herself. Like her Western sisters, Fang's fate was a terrible reminder of the ways early women alchemists suffered for their art and science. Therefore I conclude this work hoping that I have at least given voice to a few brave women's stories. There is still much to be done.

Notes

1. John R. Van Eenwyk, *Archetypes and Strange Attractors: The Chaotic World of Symbols* (Toronto: Inner City Books, 1997), 64-65.

2. Edward Grant, *The Foundations of Modern Science in the Middle Ages* (Cambridge: Cambridge University Press, 1996).

3. Allen G. Debus, *Man and Nature in the Renaissance* (Cambridge: Cambridge University Press, 1978).

4. Amird Aczel, *The Riddle of the Compass* (San Diego, Harvest, 2001).

5. Debus, 1978.

6. Debus, 70.

7. Edward Grant, *The Foundations of Modern Science in the Middle Ages* (Cambridge: Cambridge University Press, 1996).

8. Debus, 95.

9. William Bray , ed.,, *Memoirs of John Evelyn,* Volumes 1-4 (London: Henry Colburn, 1827) vol. 2, 27-28.

10. Bray, 45-46.

11. Austin Dobson, *The Diary of John Evelyn,* Volume III (London: Macmillan, 1906), 370.

12 Hall (1949).

13. Frans Wittemans, *A New and Authentic History of the Rosicrucians*. Francis Graem Davis, trans. (London: Rider and Co.; Chicago: Aries Press, 1938), 218.

14 Waite, 403.

15. Charles R. N. Routh, *Who's Who in Tudor England* (Chicago: St. James Press, 1990).

16. Hunter (1997b), xxv.

17. Edinger (1985), 183.

18. F. David Peat, *Blackfoot Physics*. (Boston: Weiser Books, 2002).

19. Van Eenwyk, 62-65.

20. Recall that the shadow represents the parts of ourselves that have been repressed or pushed into the unconscious, particularly attributes we do not wish to accept in ourselves such as prejudice or greed. Light shadow represents positive attributes that we have difficulty accepting as part of our persona.

21. Christian Bernard, F. R. C. ed., *Rosicrucian Order AMORC: Questions and Answers* (San Jose, CA: Supreme Grand Lodge of AMORC, Inc., 1996/2001).

22. Jung (1963/1970), 462.

23. Edinger (1994), 12.

24. Edinger, 12.

25. Edinger, 13.

26. For more about the collective unconscious the reader will find numerous books that address its nature as well as in the *Collected Works* of C.G. Jung.

27. Hillman (1975), 33.

28. Hillman, 35.

29. Veronica Goodchild, *Eros and Chaos: The Sacred Mysteries and Dark Shadows of Love* (York Beach, ME: Nicolas-Hays, 2001), 141-142.

30. A diary entry from an unpublished note book. See Atwood, Mary Anne. *A Suggestive Inquiry into the Hermetic Mystery with a Dissertation on the More Celebrated of the Alchemical Philosophers Being an Attempt Towards the Recovery of the Ancient Experiment of Nature* (Belfast: William Tait, 1920).

31. Arthur Cottie Burland, *The Arts of the Alchemists* (New York: Macmillan, 1967).

32. Burland, 127.

33. Burland, 128.

34. Atwood, 13.

35. Atwood, 561.

36. Atwood, 564.

37. F. David Peat, *Blackfoot Physics* (Boston: Weiser Books, 2002).

38 The exhibition moved next to Quebec and will be in Mexico City Sep-Dec, 2012.

39. Ilene Susan Fort and Tere Arcq, Terri Geis and Dawn Ades, eds., *In Wonderland: The Surrealist Adventures of Women Artists in Mexico and the United States* (Munich, Del Monico, 2012).

40. Jane Turner, ed., *The Dictionary of Art* (New York: Grove, 1996). volume 5, page 882 and *Benezit Dictionary of Artists* (Paris, Gründ, 2006).

41. Turner, 882.

42. Turner (volume 32), 5 and *Benezit Dictionary of Artists* (Volume 14), 44-45.

43. Turner, 5.

44. Gloria F. Orenstein, "The Many Worlds of Leonora Carrington: Navigating Shamanic Journeys and Surrealist Border Crossings" (presentation, ACLA conference in Puebla, Mexico, Downloaded on June 26, 2012 from: tetworld.tripod.com/gloriaorenstein.html), 3.

45. Orenstein, 3

46. Orenstein, 3.

47. Franz Dornseiff. *Das Alphabet in Mystik und Magie* (Leipzig: Verlag und Druch Von B.G. Teubner, 1922), 85.

48. Orenstein, 2.

49. Jean Chevalier and Alain Gheerbrant, *A Dictionary of Symbols*, John Buchanan-Brown, trans. (London: Penguin, 1969/1994), 150-151.

50. Orenstein, 8.

51. Orenstein, 8.

52. Edinger, 211.

53. Jung (1963/1970), 465.

54. Storr (1983).

55. Jung, 464.

56. Jung, 464.

57. Edinger, 211.

58. Edinger, 215.

59. Edinger, 218.

60. Hillman, 118.

61. Hillman, 120.

62. Romanyshyn (2002).

63. Romanyshyn, 99.

64. Romanyshyn, 104.

65. Patricia Arah Ann Taylor, "The Archetype of Enough at Threshold and in Dialogue." PhD dissertation, Pacifica Graduate Institute, 2007.

66 Jung (1960b).

67 Jung (1960), 326.

68. Woolf, 90.

69. Marelene Rayner-Canham and Geoffrey Rayner-Canham, *Women in Chemistry: Their Changing Roles from Alchemical Times to the Mid-Twentieth Century* (Philadelphia: American Chemical Society and the Chemical Heritage Foundation, 1998).

Bibliography

Abraham, Lyndy. *Alchemical Imagery.* Cambridge: Cambridge University Press, 1998.

Aczel, Amird. *The Riddle of the Compass.* San Diego, Harvest, 2001.

Agrippa, Henry Cornelius. *Fourth Book of Occult Philosophy.* Translated by Robert Turner. London: Askin Publishers, 1978/1655.

Albertus, Frater. *The Alchemist's Handbook: Manual for Practical Laboratory Alchemy.* York Beach, Maine: Samuel Weiser, 1974.

Alic, Margaret. *Hypatia's Heritage: a History of Women in Science from Antiquity Though the Nineteenth Century.* Boston: Beacon, 1986.

Andrade, Edward N. da Costa. *A Brief History of the Royal Society.* London: The Royal Society, 1960.

Atwood, Mary Anne. *A Suggestive Inquiry into the Hermetic Mystery with a Dissertation on the More Celebrated of the Alchemical Philosophers Being an Attempt Towards the Recovery of the Ancient Experiment of Nature.* Belfast: William Tait, 1920.

Batten, J.M. *John Dury: Advocate of Christian Reunion.* Chicago: University of Chicago Press, 1944.

Bayer, Penny, "Lady Margaret Clifford's Alchemical Receipt Book and the John Dee Circle." *Ambix* 52, (number 3, 2005): 271-284.

Bayley, Harold. *The Lost Language of Symbolism.* Volumes 1-2. London: Williams and Norgate, 1912.

Bernard, Christian F. R. C. ed., *Rosicrucian Order AMORC: Questions and Answers.* San Jose, CA: Supreme Grand Lodge of AMORC, Inc., 1996/2001.

Bernstein, Jerome S. *Living in the Borderland: The Evolution of Consciousness and the Challenge of Healing Trauma.* London: Routledge, 2005.

Birch, Thomas. *The History of the Royal Society of London for Improving of Natural Knowledge From its First Rise, Volume 2.* New York: Johnson Reprint Corporation, 1968.

Boas, Marie. *Robert Boyle and Seventeenth-Century Chemistry.* Cambridge: Cambridge University Press, 1958.

Borysenko, Joan. *A Woman's Journey to God.* New York: Riverhead Books, 1999.

Boyle, Robert. *Experiments and Notes about the Producibleness of Chymicall Principles being Parts of an Appendix, Designed to be Added to the Sceptical Chemist.* Oxford: H. Hall for Ric. Davis, 1680.

Bray, William ed., *Memoirs of John Evelyn, Volumes 1-4.* London: Henry Colburn, 1827.

Braybrooke, Richard Lord, ed. *Memoirs of Samuel Pepys, Esq.F.R.S.: Secretary to the Admiralty in the Reigns of Charles II. And James II. His Diary from 1659 - 1669 (2nd Ed.)* Volume 3. London: Henry Colburn, 1828.

Burland, Cottie Arthur. *The Arts of the Alchemists.* New York: Macmillan, 1967.

Capp, Bernard S. *The Fifth Monarchy Men: A Study in Seventeenth-century English Millenarianism.* London: Faber & Faber, 1972.

Case, Paul Foster. *The True and Invisible Rosicrucian Order: An Interpretation of the Rosicrucian Allegory and an Explanation of the Ten Rosicrucian Grades.* York Beach, ME: Samuel Weiser, 1981.

Cavendish, Margaret. *The Philosophical and Physical Opinions, written by her Excellency, the Lady Marchionesse of Newcastle.* London: J. Martin and J. Allestrye at the Bell in St. Pauls Church-Yard, 1655.

————. *The Life Of William Cavendish Duke of Newcastle To Which Is Added The True Relation of My Birth, Breeding, And Life.* In C. H. Firth ed., London: John C. Nimmo, 1886. (Original work published 1667)

Chevalier, Jean and Gheerbrant, Alain. *A Dictionary of Symbols,* John Buchanan-Brown, trans. London: Penguin, 1969/1994.

Chikashige, Masumi. *Alchemy and Other Chemical Achievements of the Ancient Orient: The Civilization of Japan and China in Early Times as See from the Chemical Point of View.* Tokyo: Rokakuho Uchida, 1936.

Christianson, John Robert. *On Tycho's Island: Tycho Brahe and His Assistants, 1570 – 1601.* Cambridge: Cambridge University Press, 2000.

Clifford, Margaret. *Receipts of Lady Margaret Wife of George, 3rd Earl of Cumberland for Elixirs, Tinctures, Electuaries, Cordials, Waters, etc., MS circa 1550 with Her Annotations,* 1550.

Cohn, Norman. *The Pursuit of the Millennium: Revolutionary Millenarians and Mystical Anarchists of the Middle Ages.* New York: Oxford University Press, 1970.

Considine, John in *Oxford Dictionary of National Biography,* Volume 23, edited by Mathew, H. C. G. & Harrison, B. Oxford: Oxford University Press, 2004, 833–834.

Cook, Alan, F.R.S. "Ladies in the Scientific Revolution." *Notes and Records of the Royal Society,* 51 (January 1997), 1–12.

Cook, Chris and Wroughton, John. *English Historical Facts 1603-1688.* London: Macmillan, 1980.

Copeman, William Sidney Charles. *Doctors and Disease in Tudor Times.* London: Dawsons of Pall Mall, 1960.

Cortese, Isabella. *I secreti ne'qvali si contengono cose minerali, medicinali, arteficiose.* Venice: Appresso Giouanni Bariletto, 1565. Replica printed by Book Renaissance.

Costello, Louisa Stuart. *Memoirs of Eminent Englishwomen.* London: Richard Bentley, New Burlington Street. Publisher in Ordinary to Her Majesty, 1844.

Crawford, Patricia. "Women's Published Writing 1600 – 1700." In *Women in English Society 1500 – 1800,* edited by Mary Prior, 211 - 264. London: Methuen, 1985.

Debus, Allen G. *Man and Nature in the Renaissance.* Cambridge: Cambridge University Press, 1978.

Debus, Allen, G. ed., *Science, Medicine and Society in the Renaissance.* New York: Neale Watson Academic Publications, 1972.

Debus, Allen G. & Multhauf, Robert P. *Alchemy and Chemistry in the Seventeenth Century.* Los Angeles: William Andrews Clark Memorial Library, UCLA, 1966.

Dee, John. *Monas Hieroglyphica.* London: Guliel Slivius, 1564.

Dee, John. *Private Diary of Dr. John Dee and the Catalogue of His Library of Manuscripts,* Edited by James Orchard Halliwell. London: Bowers, Nichols & Son for the Camden Society, M.DCCC.XLII.

Dick, Oliver Lawson, ed., *Aubrey's Brief Lives.* London: Secker and Warburg, 1949.

Digby, Kenelm. *Choice and Experimental Receipts in Physick and Chigery, also Cordial and Distilled Waters and Spirits, Perfumes, and other Curiosities*, 2nd ed. London: Andrew Clark, for Henry Brome at the West-End of St. Pauls, 1675.

Dobbs, Betty Jo Teeter. *The Foundation of Newton's Alchemy or "The Hunting of the Green Lion."* London: Cambridge University Press, 1975.

Dobson, Austin. *The Diary of John Evelyn*, Volumes 1-IV. London: Macmillan, 1906.

Dornseiff, Franz. *Das Alphabet in Mystik und Magie*. Leipzig: Verlag und Druch Von B.G. Teubner, 1922.

Dzielska, Maria, *Hypatia of Alexandria*. Cambridge: Harvard University Press, 1995.

Eamon, William. *The Professor of Secrets: Mystery, Medicine, and Alchemy in Renaissance Italy*. Washington, D.C.: National Geographic, 2010.

Edinger, Edward F. *Archetype of the Apocalypse: Divine Vengeance, Terrorism, and the End of the World*. Chicago: Open Court, 1999.

———. *Anatomy of the Psyche: Alchemical Symbolism in Psychotherapy*. Chicago: Open Court, 1985.

———. *The Mystery of the Coniunctio: Alchemical Image of Individuation*. Toronto: Inner City Books, 1994.

Endrei, Walter. *Old Chemical Symbols*. Budapest: Hungarian Academy of Arts and Crafts, 1974.

Epstein, Perle. *Kabbalah: The Way of the Jewish Mystic*. Boston: Shambhala, 1988.

Federmann, Reinhard. *The Royal Art of Alchemy*. R. H.Weber, trans. Philadelphia: Chilton Book Co., 1964.

Fell Smith, Charlotte. *Mary Rich, Countess of Warwick (1625 - 1678): Her Family & Friends*. London: Longmans, Green, and Co., 1901.

Fernando, Diana. *Alchemy: An Illustrated A to Z*. London: Blandford, 1998.

Firth, Colin H. ed., *The Life of the Duke and Duchess of Newcastle*. In Fritze, R. H., and Robison, W. B. eds., *Historical Dictionary of Stuart England, 1603 - 1689*. Westport, CT: Greenwood Press, 1996).

Fisher, Mitchell Salem. *Robert Boyle: Devout Naturalist, A Study in Science and Religion in the Seventeenth Century*. Philadelphia: Oshiver Studio Press, 1945.

French, Peter J. *John Dee: The World of an Elizabethan Magus*. London: Routledge and Kegan Paul, 1972.

Fritze, R. H., & Robison, W. B., eds., *Historical Dictionary of Stuart England, 1603-1689*. Westport, CT: Greenwood Press, 1996.

Gade, John Allyne. *The Life and Times of Tycho Brahe*. Princeton: Princeton University Press, 1947.

Gardiner, Dorothy ed., *The Oxinden and Peyton Letters 1642–1670*. London: Sheldon Press, 1937.

Gettings, Fred. *Dictionary of Occult, Hermetic and Alchemical Sigils*. London: Routledge and Kegan Paul, 1981.

Gibbs, Hon. Vicary, ed., *The Complete Peerage of England, Scotland, Ireland and Great Britain and the United Kingdom*, Volume 10. London: St. Catherine Press, 1910.

Gilligan, Carol. *In a Different Voice*. Cambridge: Harvard University Press, 1982/1993.

Gillispie, Charles C., ed., *Dictionary of Scientific Biography*, Volume II. New York: Charles Scribner's Sons, 1973.

Godwin, Joscelyn. *Robert Fludd: Hermetic Philosopher and Surveyor of Two Worlds*. Boston: Shambhala, 1979.

Goethe, Johann Wolfgang von. *Scientific Studies*. Translated and edited by Douglas Miller. New York: Suhrkamp, 1988.

Goodchild, Veronica. *Eros and Chaos: The Sacred Mysteries and Dark Shadows of Love*. York Beach, ME: Nicolas-Hays, 2001.

Gordon, Robin L. "The Murder of Spinoza and Other 17th Century Alchemists: A Contemporary Look at a Long-Ago Mortificatio Tale." PhD dissertation, Pacifica Graduate Institute, 2004.

Goulding, Richard N. *Margaret (Lucas) Duchess of Newcastle*. Lincoln: Lincolnshire Chronicle, 1925.

Grant, Edward. *The Foundations of Modern Science in the Middle Ages*. Cambridge: Cambridge University Press, 1996.

Gray, Ronald D. *Goethe the Alchemist*. Cambridge: Cambridge University Press, 1954. Reprinted 2002 Mansfield Centre, CT: Martino Publishing.

Greengrass, Mark, Michale Leslie, and Timothy Raylor eds., *Samuel Hartlib and Universal Reformation*. Cambridge: Cambridge University Press, 1994.

Gullan-Whur, M., *Within Reason: a Life of Spinoza*. New York: St. Martin's Press, 1998.

Halevi, Z'ev Ben Shimon. *The Way of Kabbalah*. York Beach, ME: Samuel Weiser, 1976.

———. *A Kabbalistic Universe*. York Beach, ME: Samuel Weiser, 1977.

Hall, A. Rupert., and Hall, Marie B. *The History of the Royal Society of London*, Volumes 1-4. New York: Johnson Reprint Corp, 1968.

———. *The Correspondence of Henry Oldenburg, Vol. 1*. Madison: University of Wisconsin Press, 1965.

———. *The Correspondence of Henry Oldenburg, Vol. 2*. Madison: University of Wisconsin Press, 1966.

———. *The Correspondence of Henry Oldenburg, Vol. 12*. London: Taylor and Francis, 1986.

Hall, Manly P. *Orders of Universal Reformation: Utopia*. Los Angeles: The Philosophical Society, 1949.

Hanegraaff, Wouter J. *Dictionary of Gnosis and Western Esotericism*, Volumes 1 and 2. Leiden: Brill, 2005.

Hannah, Barbara. *Jung: His Life and Work*. Wilmette, Ill: Chiron, 1997.

Hartley, Harold, F.R.S. *The Royal Society: Its Origins and Founders* London: The Royal Society, 1960.

Haydn, Joseph and Ockerby, Horace. *The Book of Dignities*, 3rd ed, Volumes 1 and 2. London: W. H. Allen, 1894.

Hillman, James. *Re-visioning Psychology*. New York: HarperCollins, 1975.

———. *A Blue Fire*. New York: Harper Perennial, 1989.

———. *The Soul's Code*. New York: Time Warner, 1996.

Hunter, Michael E., Clericuzio, Antonio., and Principe, Lawrence M. eds., *The Correspondence of Robert Boyle*, Volumes 1-6. London: Pickering and Chatto, 2001.

Hunter, Lynette. "Sisters of the Royal Society: The Circle of Katherine Jones, Lady Ranelagh." In *Women, Science and Medicine 1500-1700: Mothers and Sisters of the Royal Society*, Lynette Hunter and Sarah Hutton, eds. Gloucestershire: Sutton, 1997.

———. *The Letters of Dorothy Moore, 1612-64: the Friendships, Marriage and Intellectual Life of a Seventeenth-century Woman*. Aldershot, Hants: Ashgate, 2004.

Hunter, Lynette and Hutton, Sarah, eds., *Women, Science and Medicine 1500–1700: Mothers and Sisters of the Royal Society*. Gloucestershire: Sutton, 1997.

Hutton, Sarah, ed., *The Conway Letters: the Correspondence of Anne, Viscountess Conway, Henry More, and Their Friends 1642–1684*. Oxford: Clarendon Press, 1992.

————. *Anne Conway: a Woman Philosopher*. Cambridge: Cambridge University Press, 2004.

Israel, Jonathan I. *Radical Enlightenment: Philosophy and the Making of Modernity 1650-1750*. Oxford: Oxford University Press, 2001.

Jacob, James R. *The Scientific Revolution*. New York: Humanity, 1998.

Jotischky, Andrew and Hull, Caroline. *The Penguin Historical Atlas of the Medieval World*. London: Penguin, 2005.

Jensen, Minna Skafte. *Friendship and Poetry: Studies in Danish Neo-Latin Literature*, Marianne Pade, Karen Skovgaard-Petersen, and Peter Seeberg, eds., Copenhagen: Museum Tusculanum Press, 2004.

Josten, C. H., ed., *Elias Ashmole (1617-1692): His autobiographical and historical notes, his correspondence, and other contemporary sources relating to his life and work*. Oxford: Clarendon Press, 1966.

Jung, Carl G. "Spirit and life." In *the Structure and Dynamics of the Psyche*. Princeton, NJ: Princeton University Press, 1960a: 319-337.

————. *Alchemical Studies*. Princeton, NJ: Princeton University Press, 1967.

————. *Psychology and Alchemy*, 2nd ed. Princeton, NJ: Princeton University Press, 1953/1968.

————. *Aion, 2nd ed.* Princeton, NJ: Princeton University Press, 1959/1969.

————. *Answer to Job*, translated by R.F.C. Hull. Princeton: Princeton University Press, 1958/1969.

————. *Mysterium Coniunctionis, 2nd ed.* Princeton, NJ: Princeton University Press, 1963/1970.

————. *Memories, Dreams, Reflections*. New York: Vintage Books, 1961.

Junius, Manfred M. *The Practical Handbook of Plant Alchemy*. Rochester, VT: Healing Arts Press, 1979.

Kendal, Cumbrian Record Office, WD Hoth Box 44. *Receipts of Lady Margaret Wife of George, 3r Earl of Cumberland for Elixirs, Tinctures, Electuaries, Cordials, Waters, etc., MS circa (1550) with Her Annotations.*

Kent, Elizabeth Grey. *A Choice Manuall, or Rare and Select Secrets in Physick and Chyrurgery: Collected, and Practiced by the Right Honourable, the Countesse of Kent, Late Deceased, 2nd Ed.* London: G.D. and are to be sold by William Shears, at the sign of the Bible in St. Pauls Churchyard, 1653.

Khunrath, Heinrich. *Amphiteatrum Sapientiae Aeternae*. Hamburg: s.n., 1595.

Kingsland, William. *The Physics of the Secret Doctrine*. London: Theosophica Publishing Society, 1910.

Kirkingius, Theodore. *Basil Valentine: His Triumphant Chariot of Antimony, with Annotations of Theodore Kirkingius, M.D.* London, Dorman Newman at the Kings Arms in the Poultry, 1678.

Knoppers, Laura Lunger. *Constructing Cromwell: Ceremony, Portrait, and Print 1645-1661*. Cambridge: Cambridge University Press, 2000.

La Roche, Rebecca. *Medical Authority and Englishwomen's Herbal texts, 1550-1650*. Burlington, VT: Ashgate, 2009.

Laycock, Donald C. *The Complete Enochian Dictionary: A Dictionary of the Angelic Language as Revealed to Dr John Dee and Edward Kelley.* London, Askin, 1978.

Lea, Henry C. *A History of the Inquisition of the Middle Ages, 1-3.* New York: Harper and Brothers, 1888.

Lovell, William. *The Dukes Desk Newly Broken Up—wherin is discovered divers rare receipts of physick and surgery, good for men, women, and children.* London, 1661.

Oliver, Dick, ed., *Aubrey's Brief Lives.* London: Secker & Warburg, 1949.

MacCarthy, Bridget G. *Women Writers: Their Contribution to the English Novel 1621–1744,* Volume 1. Eire: Cork University Press, 1944.

Maddison R.E.W. *The Life of the Honourable Robert Boyle F.R.S.* London: Taylor and Francis, 1969.

Maier, Michael. *A Subtle Allegory Concerning the Secrets of Alchemy Very Useful to Possess and Pleasant to Read.* Edmonds, WA: The Alchemical Press, 1984. Reprint; Lynn Thorndike. *A History of Magic and Experimental Science, volumes VII and VIII: The Seventeenth Century.* New York: Columbia University Press, 1958.

Mathew, H. C. G., & Harrison, Brian. *Oxford Dictionary of National Biography.* Oxford: Oxford University Press, 2004.

McAllester-Jones, Mary. *Gaston Bachelard, Subversive Humanist.* Madison, WI: University of Wisconsin Press, 1991.

McIntosh, Christopher. *The Rosicrucians: The History, Mythology, and Rituals of the Esoteric Order.* York Beach, MA: Samuel Weiser, 1997.

McLean, Allen. *A Commentary on the Mutus Liber.* Grand Rapids, MI: Phanes Press, 1991.

Mede, Joseph. *Clavis Apocalyptica (English) or The key of the Revelation, searched and demonstrated out of the naturall and proper charecters of the visions. With a coment thereupon, according to the rule of the same key, published in Latine by the profoundly learned Master Joseph Mede B.D. late fellow of Christs College in Cambridge, for their use to whom God hath given a love and desire of knowing and searching into that admirable prophecie.* Translated into English by Richard More of Linley in the Countie of Salop. Esquire, one of the Burgesses in this present convention of Parliament. With a praeface written by Dr Twisse now prolocutor in the present Assembly of Divines. Imprint Printed at London: by R.B. for Phil. Stephens, at his shop in Pauls Church-yard at the signe of the gilded Lion, 1643.

Mendelson, Sara Heller. "Stuart Women's Diaries and Occasional Memoirs." In *Women in English Society 1500 – 1800,* edited by Mary Prior. London: Methuen, 1985: 181-210.

———. *The Mental World of Stuart Women: Three Studies.* Amherst: University of Massachusetts Press, 1987.

Meurdrac, Marie. *La Chymie charitable et facile, en faveur des Dames Charitable [Easy Chemistry for Women],* 1666. Paris: CNRS Editions, 1999.

Meyer, Ruth. *Clio's Circle.* New Orleans: Spring Journal, 2007.

Mlodinow, Leonard. *Euclid's Window: the Story of Geometry from Parallel Lines to Hyperspace.* New York: Touchstone, 2001.

———. *Feynman's Rainbow: A Search for Beauty in Physics and in Life.* New York: Warner Books, 2003.

Moody, Johanna ed., *The Private Life of an Elizabethan Lady: The Diary of Lady Margaret Hoby 1599-1605.* Great Britain: Sutton, 1998.

Moustakas, Clark. *Heuristic Research: Design, Methodology, and Applications.* Thousand Oaks, CA: Sage, 1990.

———. *Phenomenological Research Methods.* Thousand Oaks, CA: Sage, 1994.

Multhauf, Robert, P. *The Origins of Chemistry.* New York: Franklin Watts, 1966.

Nadler, Steven. *Spinoza: A Life.* Cambridge: Cambridge University Press, 1999.

Néroman, D. ed., *Grande Encyclopédie Illustrée des Sciences Occultesm.* Strasbourg: Editorial Argentor, 1937.

Newcastle, Margaret. *The True Relations of My Birth, Breeding, and Life.* In *The Life of William Cavendish Duke of Newcastle.* C. H. Firth, ed. London: John C. Nimmo,1886. Original work first published in 1667.

Newman, William R. In *Samuel Hartlib and Universal Reformation.* Mark Greengrass, Michael Leslie, and Timothy Raylor, eds. Cambridge: Cambridge University Press, 1994.

Nicolson, Marjorie Hope, ed. and Hutton, Sarah, revised ed., *The Conway Letters: The Correspondence of Anne, Viscountess Conway, Henry More, and their Friends 1642-1684.* Oxford: Clarendon Press, 1912.

Nummedal, Tara E., "Alchemical Reproduction and the Career of Anna Maria Zieglerin. *Ambix,* 48 (July 2001): 56-68.

Ogilvie, Marilyn, B. *Women in Science: Antiquity Through the Nineteenth Century.* Cambridge, MA: MIT Press, 1986.

Ogilvie, Marilyn and Harvey, Joy., eds., *The Biographical Dictionary of Women in Science,* Volume 1. New York: Routledge, 2000.

Oliver, Dick, ed., *Aubrey's Brief Lives.* London: Secker and Warburg, 1949.

Oliver, W. H. *Prophets and Millennialists: The Use of Biblical Prophecy in England from the 1790s to the 1840s.* Auckland: University of Auckland Press, 1978.

O'Neill, Eileen, ed., *Margaret Cavendish, Duchess of Newcastle: Observations Upon Experimental Philosophy.* Cambridge: Cambridge University Press, 2001.

Orenstein, Gloria F. "The Many Worlds of Leonora Carrington: Navigating Shamanic Journeys and Surrealist Border Crossings." American Comparative Literature Association. Downloaded on June 26, 2012 from tetworld.tripod.com/gloriaorenstein.html

Oster, Malcolm. "Millenarianism and the New Science: The Case of Robert Boyle." In *Samuel Hartlib and Universal Reformation.* Mark Greengrass, Michale Leslie, and Timothy Raylor, eds. Cambridge: Cambridge University Press, 1994: 137 - 148.

Oxen, J. H., trans. *Paracelsus. His Archidoxis, Chief Teachings Comprised in Ten Books, Disclosing the Genius Way of Making Quintessences, arcanums, magisteries, elixirs, etc.* London: Lodowick Lloyd, 1663.

Paracelsus. *His Archidoxis or, Chief Teachings; Compiled in Ten Books, Disclosing the Genunie Way of Making Quintessences, Arcanums, Magisteries, Elixirs, &c. Englished by J. H. Oxon.* London: Lodowick Lloyd, 1663.

Patai, Ralph. *The Jewish Alchemists: A History and Source Book.* Princeton, NJ: Princeton University Press, 1994.

Penney, Norman, ed., *The Journal of George Fox.* Cambridge: Cambridge University Press, 1911.

Peat, F. David. *Blackfoot Physics.* Boston: Weiser Books, 2002.

Popkin, Richard. H. "Hartlib, Dury, and the Jews." In Mark Greengrass, Michale Leslie, and Timothy Raylor, eds., *Samuel Hartlib and Universal Reformation* (Cambridge: Cambridge University Press, 1994: 118 - 136.

Porta, Giambattista della. *Natural Magick by John Baptista Porta, A Neapolitane: In Twenty Books*. London: Thomas Young and Samuel Speed, 1658.

Principe, Lawrence. In *Dictionary of Gnosis and Western Esotericism*, Volumes 1and 2. Hanegraaff, Wouter J. Leiden: Brill, 2005.

Prior, Mary, ed., *Women in English Society 1500–1800*. London: Methuen, 1985.

Queen's Closet, The. London: Obdiah Blagrave at the Sign of the Black Bear in St. Pauls Church-yard, 1683.

Raff, Jeffrey. *Jung and the Alchemical Imagination*.York Beach, ME: Nicolas-Hays, 2000.

———. *The Wedding of Sophia: The Divine Feminine in Psychoidal Alchemy*. Berwick, MN: Nicolas-Hays, 2003.

Ranelagh, Lady Katherine. *Lady Ranelaghs Medical Reciepts*. Sloane 1367.

———. *Kitchen-Physick*. Wellcome Medical Science 1340.

Rattansi P. M. "Newton's Alchemical Studies." In *Science, Medicine and Society in the Renaissance,* edited by Allen Debus, Allen, G. New York: Neale Watson Academic Publications, 1972.

Rayner-Canham, Marelene and Rayner-Canham, Geoffrey. *Women in Chemistry: Their Changing Roles from Alchemical Times to the Mid-Twentieth Century*. Philadelphia: American Chemical Society and the Chemical Heritage Foundation, 1998.

Read, J. *Prelude to Chemistry: An Outline of Alchemy, Its Literature and Relationships*. New York: Macmillan, 1937.

Redgrove, H. Stanley. *Alchemy: Ancient and Modern*. New Hyde Park, NY: University Books, 1969.

Riddell, Edwin, ed., *Lives of the Stuart age 1603 - 1714*. New York: Harper and Row, 1976.

Roach, Christy A., trans., *La Chymie charitable et facile, en faveur des Dames Charitable[Easy Chemistry for Women]*, 1666. (Unpublished translation)

Robertson, William. *The First Gate, Or, The Outward Door to the Holy Tongue, Opened in English*. London: Evan Tyler, for Humphrey Robinson, at the three Pigeons in St. Paul's Churchyard,1654.

———. *The Second Gate, or, The Inner Door to the Holy Tongue*. London: Evan Tyler, for Humphrey Robinson, at the three Pigeons in St. Paul's Churchyard, 1655.

Robinson, Henry W. and Adams, Walter. *The Diary of Robert Hooke, M.A., M.D., F.R.S. 1672-1680*. London: Taylor and Francis, 1935.

Rogers, Philip G. *The Fifth Monarchy Men*. London: Oxford University Press, 1966.

Ronnberg, Ami and Martin, Kathleen, eds., *The Book of Symbols*. Cologne, Germany: Taschen, 2010.

Romanyshyn, Robert D. *Mirror and Metaphor: Images and Stories of Psychological Life*. Pittsburg: Trivium, 2001/1982.

Routh, Charles R. N. *Who's Who in Tudor England*. Chicago: St. James Press, 1990.

Rubin, Stanley. *Medieval English Medicine*. London: David and Charles, 1974.

Rulandus, Martinus. *A Lexicon of Alchemy Or Alchemical Dictionary*. Frankfurt: Zachariah Palthenus, 1612 - Reprinted by Montana, U.S.A.: Kessinger Publishing.

Russell, Dora. *Hypatia: Or Woman of Knowledge*. New York: Dutton, 1925.

Sackville-West, V. *The Diary of the Lady Anne Clifford with an Introductory Note by V. Sackville-West.* London: William Heinemann, 1923.

Schneider, Wolfgang. "Chemiatry and Iatrochemistry." In *Science, Medicine and Society in the Renaissance.* Allen Debus, ed. New York: Neale Watson Academic Publications, 1972.

Sendivogius, Michael. *The New Chemical Light,* n.d.

Shakespeare, William. In Wright, L. B. and LaMar, V. A., eds., *As You like It* New York: Washington Square Press, 1968.

Sigerist, Henry E., ed., *Paracelsus: Four Treatises.* Baltimore, MD: Johns Hopkins University Press, 1941.

Smith, Cyril Stanley and Hawthorne, John G., "Mappa Clavicula: An Annotated Translation Based on a Collation of the Sélestat and Phillips-Corning Manuscripts with Reproductions of the Two Manuscripts," *Transactions of the American Philosophical Society,* 64 (number 4). Philadelphia: American Philosophical Society, 1974: 3-128.

Sobel, Dava. *Galileo's Daughter.* New York: Penguin, 2000.

Stavish, Mark. *The Path of Alchemy.* Woodbury, MN: Llewellyn, 2006.

Stimson, Dorothy. *Scientists and Amateurs: A History of the Royal Society.* New York: Henry Schuman, 1948.

Stitt, Bette, ed., *Diana Astry's Recipe Book c. 1700. Bedfordshire Historical Record Society, Vol. 37.* Luton, Bedfordshire: Society at   Streatley, 1955.

Talbot, Alathea. *Natura Exenterata: Or Nature Unbowelled.* London: H. Twiford, 1655.

Taylor, Patricia Arah Ann. "The Archetype of Enough at Threshold and in Dialogue." PhD dissertation, Pacifica Graduate Institute, 2007.

Temkin, C. Lilian, trans. "Seven Defensiones, the Reply to Certain Calumniations of His enemies." In *Paracelsus: Four Treatises.* Henry E. Sigerist, eds. Baltimore, MD: Johns Hopkins University Press, 1941.

The Complete Peerage or a History of the House of Lords and All Its Members from the Earliest Times, Volume 10. London: St. Catherine Press, 1945.

Thompson, C. J. S. *Alchemy and Alchemists.* New York: Dover, 1932/2002.

Thoren, Victor E. *The Lord of Uraniborg: A Biography of Tycho Brahe.* Cambridge: Cambridge University Press, 1990.

Thorndike, Lynn. *A History of Magic and Experimental Science, volumes V and VI: The Sixteenth Century.* New York: Columbia University Press, 1958a.

———. *A History of Magic and Experimental Science, volumes VII and VIII: The Seventeenth Century.* New York: Columbia University Press, 1958b.

Tosi, Lucia, "Marie Meurdrac: Paracelsian Chemist and Feminist," *Ambix* 48 (July 2001): 69-82.

Turnbull, G. H., ed., *Letters Written by John Dury in Sweden, 1636-38.* Sheffield, England: University of Sheffield, 1950.

Turner, Jane, ed., *The Dictionary of Art.* New York: Grove, 1996.

Urbigerus, Baro, *The One Hundred Alchemical Aphorisms of Baro Urbigerus.* Edmonds, WA: The Alchemical Press, 1986/1997.

Van Eenwyk, John R. *Archetypes and Strange Attractors: The Chaotic World of Symbols.* Toronto: Inner City Books, 1997.

Von Franz, Marie Louise. *Alchemy: An Introduction to the Symbolism and the Psychology.* Toronto: Inner City Books, 1980.

Waite, Arthur E. *The Real History of the Rosicrucians.* London: George Redway, 1887.

————. *The Holy Kabbalah.* New Hyde Park, NY: University Books, 1960. (Work originally published in 1902 as *The Doctrine and Literature of the Kabbalah* with Theosophical Society in London as publisher)

Wallis, John. *A Defense of the Royal Society, and the Philosophical Transactions, particulary Those of July 1670.* London: for Thomas Moore, at the Maidenhead, 1678.

Wheatley, Henry B., ed., *The Diary of Samuel Pepys M.A. F.R.S.,* Volume 6. London: George Bell & Sons, 1900.

Whitaker, Katie. *Mad Madge: the Extraordinary Life of Margaret Cavendish, Duchess of Newcastle, the First Woman to Live by Her Pen.* New York: Basic Books, 2002.

White, Michael. *Isaac Newton: the Last Sorcerer.* Reading, MA: Perseus Books, 1997.

Wiesner, Merry E. *Women and Gender in Early Modern Europe,* 2nd Ed. Cambridge: Cambridge University Press, 2000.

Wittemans, Frans. *A New and Authentic History of the Rosicrucians.* Francis Graem Davis, trans. London: Rider and Co.; Chicago: Aries Press, 1938.

Wooley, Hannah. *The Ladies Delight: Or a Rich Closet of Choice Experiments & Curiosities.* London: I. Pilbourn, 1672.

Woolf, Virginia. *A Room of One's Own.* New York: Harcourt, 1929/1957.

Yates, Frances. *The Rosicrucian Enlightenment.* London: Routledge, 1972.

————. *The Occult Philosophy in the Elizabethan Age.* London: Routledge, 1979.

Young, Frances Berkeley. *Mary Sidney Countess of Pembroke.* London, David Nutt, 1912.

Zimmer, Carl. *Soul Made Flesh: The Discovery of the Brain—and How It Changed the World.* New York: Free Press, 2004.

Index

J

Jewish Alchemists 21, 52, 203
Jewish female Alchemist 188, 190
Jones, Katherine Boyle 3, 92-100,
    127, 159-63, 170-1
Jung, Carl G. 6, 41-7, 53-5, 190-2,
    195-6

K

Kabbalah 50-2, 130-4, 155-75, 184-6,
    188-91
Kitchin-Physick 98, 100-1

L

*Lady Margaret Clifford's Alchemical
    Receipt Book* 122, 197
Lady Ranelagh 20-1, 52, 89, 92, 97
Lady Ranelaghs Choice Receipts 98-
    100
Leonora Carrington 60, 186-8, 190,
    195
Lovell, William 110, 115

M

maggid 76-7
magic 21, 34, 101-3, 156, 204
Magnum Opus 2, 29-30, 57-63, 83, 90
Maier, Michael 30, 34-5, 54, 201
Mappae Calvicula 106-7
Margaret Manuscript 118, 122-5
masculine model 11-12
matter, plant 27, 36
Mede, Joseph 170-1, 173, 175, 201
medicine 2-3, 8-9, 32-4, 83-4, 86-7
mercury 25, 30, 35-6, 74-5, 194
metals 2, 24-5, 27-30, 59, 86
metaphor, discussed Jung's alchemical
    46
Meurdrac, Marie 1, 5, 57, 84-7, 111-
    12
millenarians 12, 97, 161-3, 165-8, 171
millennium 163, 166-7, 171, 198
minerals 1, 24-8, 36, 86, 123-4

modern science 54, 92, 102, 177, 194
monads 130-2
monarchy 158, 163, 167-70
moon 2-3, 25, 101, 146-7, 178
Moore, Dorothy 20, 98, 156-9, 162-4,
    166
mortificatio 28, 48-50, 60, 192
Moustakas, Clark 6, 21, 186
Mutus Liber 4, 21, 202

N

Natura Exenterata 88-91, 112, 117,
    204
Natural Magik 84, 108-9
natural philosophy 5, 7, 19, 32, 143-5
Newcastle, Margaret Cavendish Duch-
    ess of 138-9, 152, 197, 202
notation, typical alchemical 99-100
Nummedal, Tara 5, 9, 21, 71-6, 81

O

occult sciences 102-3
Oldenburg, Henry 158-62, 200
opposites 42, 48-50, 182, 192
opus 49, 73, 75, 190
Orenstein, Gloria 186, 188-9, 195-6
Oxford group 160
oyle 105-6, 129

P

Paracelsus 9, 32-4, 66-7, 74-5, 203-4
  followers of 177-8
pelican 73, 90
Pepys, Samuel 10, 141-2, 144-5, 160,
    197
perfection 26, 162, 172
Philosopher's Stone 4-5, 8-9, 23-5, 49-
    50, 72-3
Philosophical Child 49, 74, 172, 189,
    191-2
Philosophical Transactions 149, 157,
    159-60, 174
physical alchemy 7, 18, 32, 37, 183-4
physician-alchemist 34